Methods for the Assessment of Structural Integrity of Components and Structures

Edited by
David Lidbury and
Peter Hirsch

MANEY

FOR THE INSTITUTE OF MATERIALS, MINERALS AND MINING

B0778
First Published in 2003 for
The Institute of Materials, Minerals and Mining by
Maney Publishing
1 Carlton House Terrace
London SW1Y 5DB

ISBN 1-902653-78-5

Typeset in the UK by
Keyset Composition, Colchester

Printed and bound in the UK by
Charlesworth Group, Huddersfield

Contents

Foreword

This volume brings together the papers presented at the 8th Symposium organised by the Technical Advisory Group on Structural Integrity of Nuclear Plant (TAGSI), held on 25th April 2001 at the TWI Conference Centre at Granta Park, Great Abington near Cambridge, UK. The subject of this one-day symposium was ^Methods for the Assessment of the Structural Integrity of Components and Structures. It was attended by over eighty delegates from nuclear and non-nuclear industries, academia and Government organisations.

The construction, process, power generation, manufacturing and transport industries all rely on the integrity of appropriate structures and components to ensure the safe and economic operation of their plant. The object of the Symposium was to disseminate the various methodologies used in different industries for the assessment of structural integrity, and to permit comparisons to be made of the different approaches. There were three sessions. The first included an overview of the key elements in the assessment process leading to a safety argument, a repair, a replacement or an operational solution to a structural integrity problem, and a view by the UK Regulator on the Safety Assessment Principles in the Nuclear industry. There was another paper by the UK Regulator on the comparison of the codes used for Engineering Critical Assessments of conventional pressure plant.

The second session was concerned in part with recent progress worldwide on fracture assessment codes, including in particular an overview of the SINTAP procedure, and of the latest revision to the R6 code; another contribution discussed recent work relevant to the question of transferability of fracture toughness parameters between test specimens and structure. A paper on fatigue-based structural assessments for aircraft structures highlighted the challenges for maintenance of structural integrity brought about by the use of new materials and fabrication processes.

The third session consisted of papers by users of structural integrity codes in applications in various industries. A paper on the code requirements for aero-gas turbines described the approaches used for engine discs, blades and thin shell components such as pressure casings and combustors. The application of codes in various civil engineering structures was illustrated by examples which included *inter alia* the spindle of the London Eye, fracture requirements for steel bridges and assessment of defects in an airport roof structure. A paper on the application of Engineering Critical Assessment of off-shore structures included examples of mobile drilling units and pipelines. The contribution of the application of integrity codes to pressurised components in the process industry described the different codes utilised, and emphasised the conservative approach used in component design. The session ended with a general summary by Professor Knott of future needs for the assessment of the integrity of components and structures. The Symposium was remarkable for the great variety of structural integrity issues addressed.

P.B. Hirsch
D.P.G. Lidbury

CHAPTER 1

The Development of Structural Integrity Assessment Methods: An Overview

A. R. Dowling and P. E. J. Flewitt

BNFL Magnox Generation, Berkeley Centre, Berkeley, Gloucester GL13 9PB, UK

ABSTRACT

Understanding the integrity of critical structures has been a driving force for the development of engineering analysis methods since Galileo, in the seventeenth century, investigated the resistance of a beam subject to bending loads. Construction, process, power generation, manufacturing and transport industries need to be able to assure the safe and economic operation of their plant by relying on the integrity of appropriate structures and components. The design engineer assists in the creation of a structure or component by using analytical tools and supporting data most appropriate for the application. By comparison, the operator of the plant has to ensure that, throughout the service life, it is secure against the design intent. Both offer challenges with respect to the structural integrity diagnosis and assessment which forms an essential input to providing the necessary assurance of safe and reliable operation. In general, failure of structures or components can have profound commercial and financial implications. However, there are areas where failures are completely unacceptable due to the impact on the environment and society. Those industries where this is particularly relevant are nuclear, chemical and transportation. To specify and formulate the overall strategy for safe, reliable and economic operation requires a comprehensive understanding of the capability of the plant to meet the required duty based upon a structural integrity assessment. In this overview, we consider what assessment methods designers and assessors require, the tools available and the needs of specific industries. Recent and future developments are discussed briefly.

INTRODUCTION

Understanding the integrity of critical structures has been a driving force for the development of engineering analysis methods since Galileo, in the seventeenth century, investigated the resistance of a beam subject to bending loads.[1] Construction, process, power generation, manufacturing and transport industries need to be able to assure the safe and economic operation of their plant by relying

1

on the integrity of appropriate structures and components.[2, 3] The design engineer assists in the creation of a structure or component by using analytical tools and supporting data most appropriate for the application. By comparison, the operator of the plant has to ensure that, throughout the service life, it is secure against the design intent. Both offer challenges with respect to the structural integrity diagnosis and assessment which forms an essential input to providing the necessary assurance of safe and reliable operation.

At the present time, structural integrity assurance may be considered as the demonstration that the structure or component meets the required duty with an appropriate consideration of safety and economics. As a consequence, it is a multi-disciplinary activity[2, 4] that draws together a number of specialised fields including inspection and plant monitoring, materials science including properties on the macro, meso and micro scales, welding technology, structural analysis and general engineering safety and economic assessment. The main structural integrity inputs to a particular operational plant problem are the inspection and monitoring, diagnosis and assessment, Fig. 1. These span from plant investigation work at site to longer term research and development activities to understand and solve the problems. Inspection includes provision of the hardware for techniques together with the necessary interpretative capability to support the detection limits and the resolution for sizing defects. When diagnosing a problem, work is required to provide an understanding of the mechanisms leading to any deterioration of the plant integrity or potential failure to allow the correct structural assessment to be undertaken. These are the overall inputs to the structural integrity argument or

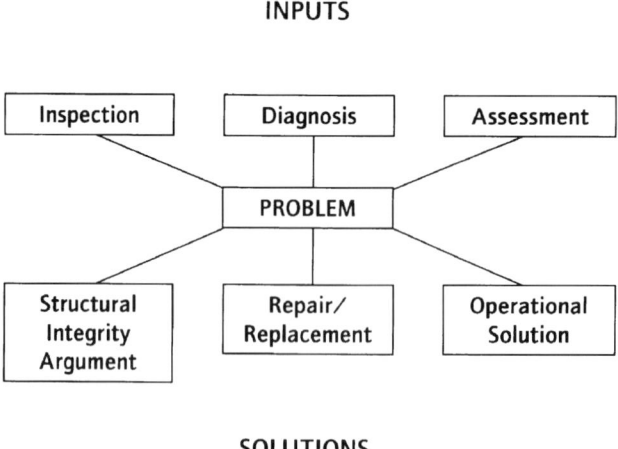

Fig. 1   The key elements for a structural integrity assessment which leads to a safety argument, a repair or replacement, or an operational solution to a problem.

safety case that would be used to underwrite continued operation, a repair or replacement strategy, or an operational solution. Equally, the designer of structures follows a guided, interactive process directed to achieving a practical balance between state-of-the-art structural capability and the intended service requirements.[5] These capabilities and requirements are typically assessed through a process comprising regulations, methods of analysis, databases and validation tests. Static design of structures and components has evolved alongside the development of widely accepted analytical tools and design procedures that reflect combined engineering experience derived from service/operational knowledge.

In general, the probability of a failure with time in service for structures and components follows a curve of the form shown in Fig. 2, where during early and late life there is a greater probability of failure.[6] Problems encountered during commissioning and early life usually result from poor design or inadequacies in fabrication, construction and assembly. Increases in the later stages of life are more often associated with degradation of the properties of materials and the onset of time-dependent failure mechanisms. At intermediate stages, problems are usually a result of the unforeseen response of the material to the service environment, specific requirements of the plant-operating regime, inadequacies in manufacturing or simply poor maintenance.[7,8] Unfortunately, there are examples where catastrophic failures have occurred, although these are rare. A particular set of failures was associated with brittle fracture of ferritic steels relating to an inadequate knowledge of the ductile to brittle transition.[9,10] These include the well-documented catastrophes associated with the liberty ships in the 1940s through the problems associated with tankers even as late as the 1970s. The failure of bridges, for example the Kings Bridge in Melbourne in 1962 has also been attributed to brittle fracture. In 1965, a pressure vessel

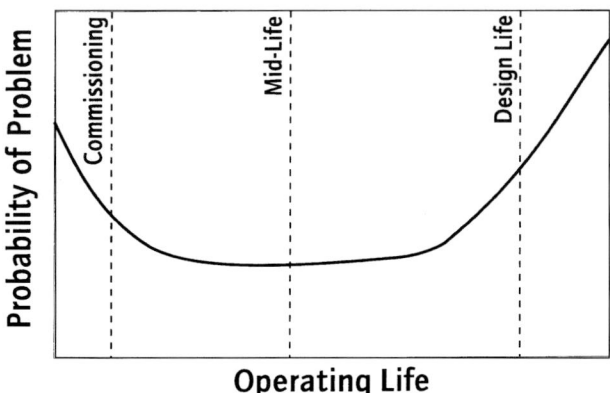

Fig. 2 The change in the probability of the incidence of structural integrity problems in relation to the operating life of plant.

fabricated from 149 mm thick steel and intended to be used in an ammonia plant in Immingham (UK) failed during proof test. There are other examples such as the failure of a low pressure stage steam turbine disc manufactured from a low alloy steel which burst in service due to a combination of inappropriate heat treatment and environmentally assisted cracking,[11, 12] Fig. 3. This was a direct result of a more onerous service environment, off-load, not addressed in the design, combined with poor materials properties. Each case leads to a reconsideration of the methods or codes on which the designs are based and appropriate revisions are introduced to take account of this operation/engineering experience.

Failure of structures or components of the types described above can have profound commercial and financial implications. However, there are areas where such events are completely unacceptable due to the impact on the environment and society. Those industries where this is particularly relevant are nuclear, chemical and transportation. Certainly in the UK the safe, reliable and economic continued operation of nuclear plant including electrical power generating plant, is assured by the use of planned maintenance schedules combined with an appropriate range of structural integrity assessments and safety arguments. To specify and formulate this overall strategy for nuclear power generating plant requires a comprehensive understanding of the capability of the plant to meet the required duty based upon a structural integrity assessment. In this overview, we consider the assessment methods that designers and assessors require, the tools available to them, and the needs of specific industries. Recent and future developments will be discussed briefly before the concluding statements.

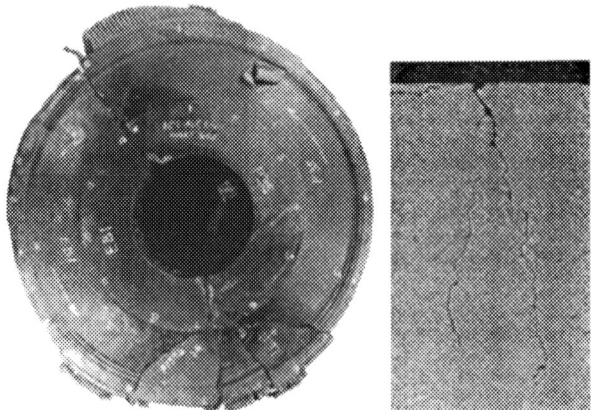

Fig. 3   Failure of a low pressure stage steam turbine disc manufactured from a low alloy ferritic steel: (a) Reassembled pieces of the 1.5 m diameter disc which failed while rotating at 50 Hz; (b) failure was caused by the growth of a small SCC crack in a steel of low fracture toughness.

METHODS DESIGNERS AND ASSESSORS REQUIRE

In general, designers and assessors of structures and components require structural integrity assessment methodology that is conservative and of appropriate accuracy to achieve a required level of safety and reliability. As shown schematically in Fig. 4 a simplified deterministic analysis adopts bounding assumptions whenever an uncertainty exists to achieve this conservatism.[13] If a simple, code-based structural integrity assessment gives a level of safety that would require, for example, a costly repair or some other action, the option remains to undertake a more detailed and rigorous analysis which seeks alleviations to the conservatisms, but does not impair the true safety. Certainly, a simplified analysis should always provide a more conservative assessment compared with a detailed analysis. However, several of the parameters used in deterministic fracture mechanics assessments are probabilistically distributed. For example, material properties exhibit scatter, defects and their sizes are distributed and loads may be random. Consequently, deterministic assessments assume bounding values for the variables to achieve conservatism. But in many cases over-conservatism is introduced. In this respect, a probabilistic approach offers the potential to include the appropriate level of detail into the analysis leading to a more realistic assessment at the prescribed level of safety. In addition, although each deterministic assessment could be undertaken in an *ad hoc* way, the use of standards, codes and defined procedures leads to greater efficiency, continuity and acceptability to the designer, operator and regulatory body and, indeed, the public.

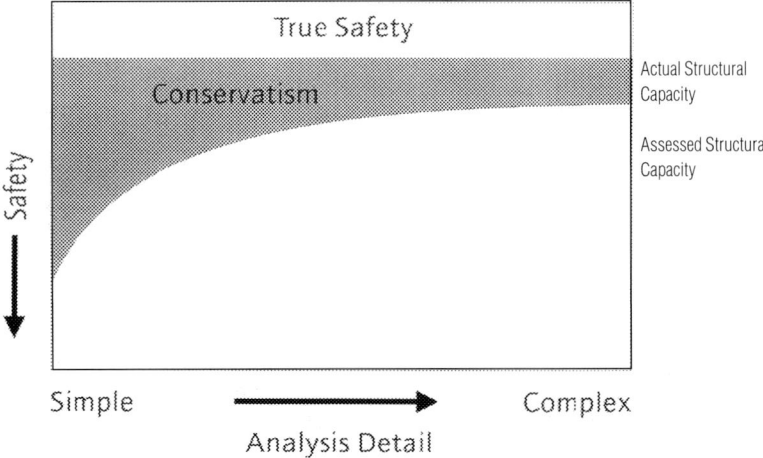

Fig. 4 Schematic diagram showing the amount of analysis detail required with respect to the level of safety associated with a structural integrity assessment.

If the required measure of reliability and safety is to be underwritten, key inputs to structural integrity assessment methods are the mechanical and physical properties of the materials selected. Reliability has to focus on the potential failure modes and how well these can be predicted. In general, diagnosis of the reasons for failure after they have occurred is more secure than prediction of service life. As a result, designers accommodate uncertainties by adding conservatisms and supporting them, particularly for high integrity plant, by use of surveillance specimens, monitoring operating conditions and regular non-destructive inspections. The approach is usually set into a declared strategy that uses these in combination. A challenge is to develop predictive models combined with mechanical property data to support the strategy. Here difficulties arise because the structures and components are usually of larger cross section than test specimens, invariably include features such as weldments, and the environment and stress state are also more complex. In addition, the period of service required usually exceeds that achievable in tests, so that acceleration in parameters including stress, temperature and environment is required combined with extrapolation to service conditions. Hence there is a need to demonstrate that the controlling mechanisms in the accelerated tests are the same as those experienced during service.

TOOLS AVAILABLE

EMPIRICAL CODES FOR ENGINEERING DESIGN

The most important contribution to the assurance of structural integrity is made at the design and construction phase. It is at this stage that previous experience and practices can and should be called upon to provide confidence in the ability of a structure or component to fulfil the design intent. In most industries this previous experience has been collected and collated in the form of design codes and standards. This practice, initiated in the nineteenth century, is now a universal feature adopted by the engineer. The principle embodied in design codes is that relatively straightforward recipe books should be available for non-specialist engineers to follow in constructing or manufacturing reliable plant. They embody previous experience and traditionally were modified when either poor performance or failures of structures and components were realised. Thus a series of boiler vessel failures in the early life of the steam age led directly to the production of the first Boiler and Pressure Vessel Codes of Practice.

Empirical design codes are clearly at their most efficient when the application exactly replicates the previously successful practice. There are two factors which threaten replication. The first is that despite all good intentions, exact replication is not achieved in practice because of divergencies such as the introduction of rogue materials, undiscovered cracks or other defects, geometrical differences and environmental effects. The second factor is that the history of technological development presents a constant desire for efficiency, performance, cost or other improvements. The boundaries of past practice are challenged and it is desirable to enter new regimes of operation, materials of construction, production methods, plant life expectations and so on. It is in these areas where the 'safe umbrella' of a design code becomes less secure. Methods of understanding and assessing the new conditions must be put in place, following which, it is possible to include the results in the design codes.

Scientific and technical developments assist in this process so that the opportunity for today's engineer to understand the fundamental behaviour of materials and to calculate the loads imposed on an engineering component is unrecognisable compared with even 30 years ago when many of today's operational structures and components were designed. These results of scientific research allow more confidence to be placed in structural integrity assurances presented for unfamiliar or unusual situations. In consequence, this allows the boundaries of empirical codes to be extended with the necessary level of confidence.

DEVELOPMENT OF STRESS ANALYSIS

Within the sphere of engineering, structures and components are made from a wide spectrum of materials. Whatever material is employed, the structural integrity assessment requires that the chosen material should not 'fail'. There are many descriptions for what 'failure' means and it depends very much on the intended purpose. For some structures and components, large scale deflection is no problem. For most, catastrophic failure is to be avoided; for others, dimensional tolerance is crucial. In other situations, holes or leakage paths cannot be tolerated. What is common to all is that knowing the loads, or more correctly, stresses applied, together with knowledge of the material response to stress, conclusions can be drawn about fitness for purpose of the structure or component of interest. A crucial part of structural integrity assurance is therefore the obtaining of accurate values, or at least over-estimates, of the stresses experienced by the plant. One of the simplest rules of successful operation is, where possible, to keep material response within the elastic region.

Very early analysis methods were only capable of calculating stresses in simple structures such as cylinders or spheres under relatively simple loadings such as pressure, global bending, or perhaps one-dimensional temperature variation. Rules of compatibility and equilibrium together with simplification of stress distributions into membrane and bending components were employed to derive stresses for more complex geometries. The introduction of girders, angles and I-beams spurred a large effort in understanding the stressing of these items leading to the production of standard handbooks for general engineering use; a slide rule was the trademark of an engineer.

Different industries followed similar practices in developing handbooks of stress analysis solutions for increasingly complex geometries, e.g. Roark.[14] The introduction of electronic calculators allowed greater freedom and ingenuity in adapting and adopting these solutions, but the advent of the digital computer brought with it a revolution in stress analysis methods, the Finite Element Method. Finite element analysis has developed rapidly over the last thirty years, so that it is now an extremely accurate tool for deriving the interior stresses in complex components subjected to all manner of loading situations. Finite element methods are widely available but nevertheless, like all computing techniques, it is important to ensure that the computed results bear a close relationship to reality; this remains a skilled job. Fundamentally, with finite element methods, there is no difference from the early stress analysis techniques described above. A structure or component is divided into a large number of simple geometrical shapes, each one of which can be readily analysed. Compatibility and equilibrium conditions for each element with its neighbours are satisfied but the number of iterations of these simple calculations demands computing power. However, there are still some important inputs, material behaviour, the applied loading and any restraints or constraints that must all be modelled adequately. The number of elements necessary to provide sufficiently accurate results is also an item which needs specialist attention. There are a number of types of analyses which can be undertaken using the finite element methodology. The simplest is linear elastic analysis but other methods of analysis, non-linear elastic, plastic, dynamic and impact can be invoked, although these tend to become progressively more difficult. Add to these the possibility of analysing structures containing defects, which present a mathematical singularity, and the subject is extensive. Industry guidelines to finite element analysis have been produced, e.g. NAFEMS,[15] and the R6 procedure[16] also includes a guidance section providing advice on mesh requirement, input requirements, user accuracy and sufficiency checks, and types of analyses which can be performed.

Finite element analysis allows very accurate estimates of stresses and material response to be calculated and thereby allows designers to design closer to safe limits. Despite the benefits, this has the potentially detrimental effect of reducing the margins that exist in a structure or component against all adverse conditions whether these arise from unplanned loadings, poor construction and manufacture, or rogue materials, etc. Another consequence of this ability to design closer to recognised limits is that the proximity to limits can be engineered throughout a component in the name of efficient design. The consequence of this approach is that when a real limit is reached the efficiently designed structure or component tends to fail catastrophically rather than search out the weak points which were inevitably present when using more primitive design methods. The advances in stress analysis have allowed enormous improvements to be made in the understanding of structural behaviour. They have stimulated fitness-for-purpose assessments to enable 'code' rules to be extended and have contributed significantly to the concept of efficient design in terms of material usage. The engineer must, however, continue to be aware of the threats such increased knowledge can bring.

RANGE OF STRUCTURAL ASSESSMENT METHODOLOGIES

It has already been stated in the section 'Empirical Codes for Engineering Design' that the most important contributions to structural safety are made at the design, manufacturing and construction stages. A fundamental part of this process is to ensure that structures and components are made with the highest possible standards of workmanship. For many decades one driver for the rules on material selection and manufacturing practices was to minimise the potential for defects, in particular crack-like defects. It was nevertheless recognised that the presence of defects could not be entirely eliminated, and the emergence of better non-destructive testing techniques demonstrated that defects could and did exist in operating plant. What is more, it was recognised that many items which had operated successfully for more than their design lives contained defects whereas some defects were capable of producing catastrophic failures e.g. Hodgson and Boyd,[9] Brown.[17] The elimination of all defects from service components is a practical impossibility and a financial non-starter. It was necessary to have available a method for distinguishing between those defects which were potentially disastrous and those which would not affect safe normal operation. The difficulties associated with crack-like defects became more apparent with the emergence of new joining technology, in particular welding. Propagation of cracks in riveted or bolted flanged joints was generally less serious because cracks were unable to 'jump' the interface. Failure of bolts was generally incremental, allowing time for observation of leaking flanges, missing bolt heads, etc. This is another example of the drive for efficiency leading to more dramatic events if failure initiated.

There was clearly a need for better assessment of defects. This was stimulated in the aerospace industry by the study of fatigue leading to the production of fatigue guidelines and guidelines on the fracture response of the thin sheets used for the external skin of aircraft. Within the heavier industries, defect assessment procedures were developed to address the emerging issues. The first attempts to produce a standard on this subject were commenced by British Standards Institution based on the crack opening displacement work at the then Welding Institute, now TWI. This work was directed at the offshore and heavy engineering industry where the as-welded structures and the use of structural welds without post-weld heat treatment was commonplace. The work relied heavily on experimental observations of the fracture resistance capabilities of artificial defects introduced into weldments with and without post weld heat treatment. This was essentially a strain-based procedure using a critical crack opening displacement as a criterion for whether cracks would propagate. The work eventually led to the production of the British Standards Published Document PD6493:1980, which included procedures for assessing fatigue and fracture of welded structures. This document was relevant to all industries.

Within the nuclear industry, there was an even more demanding requirement for the accurate assessment of whether plant could operate with defects because of the potential consequences of a wrong decision. Most relevant structures and components were stress relieved and the British Standard developments did not appear to be producing the right procedure at the right time. The Central Electricity Generating Board therefore developed its own defect assessment procedure to be known as R6. This procedure, first published in 1976,[18] had an immediate impact in the field, turning fracture mechanics from a complex, esoteric subject considered only by experts into a tool which could be used with confidence by the suitably trained engineer. The secret was the presentation of the assessment results in the form of the now familiar failure assessment diagram, Fig. 5. This gave a powerful visual indication of the proximity of a structure to the limiting condition. The value of the presentation was enhanced by the ability to demonstrate the effect that changes to contributory parameters had on an assessment. Previously, this had been possible only by performing a series of relatively complicated calculations. The R6 procedure did not provide the extensive guidance on fatigue as presented within PD6493, since fatigue was generally of minor interest in the types of structures and components operated at that time by the CEGB.

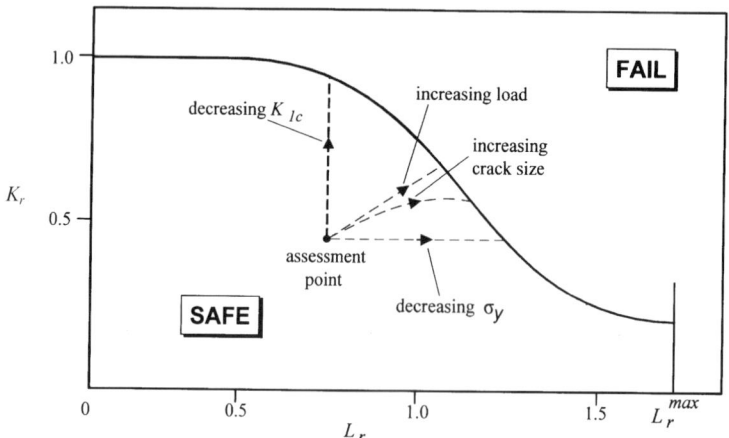

Fig. 5 A typical R6 failure assessment diagram showing an assessment point within the safe region, together with the trajectories of this point when fracture toughness K_{Ic}, load and crack size are varied.

Since that time various defect assessment methods have been developed and have grown together. There is now almost universal acceptance of the benefits of a failure assessment diagram although specific formulations differ. The crack opening displacement methods have been accommodated into the failure assessment diagram framework in BS7910:1999. In 1996, a European collaboration was funded to investigate defect assessment methodologies and charged with the objective of producing a European defect assessment methodology. This collaboration reported in 1999 and, although it relied heavily on the R6 procedure, developments and good practices from the other participants were incorporated.

MATERIAL SELECTION AND MECHANISTIC MAPS

Aspects of structural integrity assurance which are often under-emphasised are those of material selection and the underlying failure mechanisms. It is again at the design stage that these aspects should receive the greatest attention. It was stated in the section 'Empirical Codes for Engineering Design' that design codes tend to catalogue prior experience which has provided satisfactory performance. Prior experience does, however, stifle innovation, and material selection is one area where design codes may be considered to be overly restrictive. Continuing the theme of fitness-for-purpose from the last section, it is sometimes advantageous to stand back and revisit the performance specification when considering material selection.

Materials, of course, encompass a huge diversity and typical properties whether physical, mechanical, chemical or thermal also demonstrate a very broad spectrum. It seems that to truly stand back and investigate all possible materials for a particular application is a particularly demanding task. Fortunately, the idea of material property maps was introduced by Ashby,[19] allowing an easier consideration of material selection again by providing a visualisation. The idea is extremely simple in that a relevant property is plotted against another using logarithmic scales to encompass the wide variety of property values. It was observed that data for particular classes of materials clustered together on these plots, the major classes being metals, polymers, ceramics, elastomers, glasses and composites. A typical diagram, Fig. 6(a), addresses Young's modulus for elasticity, E, and density ρ. Also shown are guide lines for material selection for various conditions, maximising E/ρ indicates the material for the lightest rod to carry a load for a given deflection, maximising $E^{1/2}/\rho$ indicates the material which would provide the lightest column to withstand buckling for a given compressive load and maximising $E^{1/3}/\rho$ indicates the material to make a panel providing the minimum deflection for a given applied pressure. Probably the most important properties to use in constructing these diagrams are density, Young's modulus, strength, fracture toughness, thermal expansion, conductivity and diffusivity, damping coefficient and specific heat. The component performance requirements can be formulated in terms of these parameters and successive iterative use of a series of these property diagrams can assist the designer in selecting the most appropriate class of material and, by drilling down, the most appropriate material within the class.

As considered previously, there is a need to ensure that the mechanisms associated with accelerated test data remain those experienced in service. Hence there are benefits associated with being able to develop realistic mechanistic maps, for example deformation and fracture maps, Fig. 6(b), extending to stresses, temperatures and environments appropriate to service. Again, by use of such maps Ashby[19] was able to show, for example, the temperature and stress ranges on which particular creep deformation mechanisms apply. These are generally indicative since there is a need to accommodate specific factors such as material composition, heat treatment, microstructure, etc. which, if available, greatly assists the necessary extrapolations.

MECHANISTIC UNDERSTANDING AND LIFE PREDICTION

Design is focused on expectation, whereas safe continued operation is more concerned with reality. It is prudent in all industries to monitor the experience of critical structures and components to ensure performance continues to be

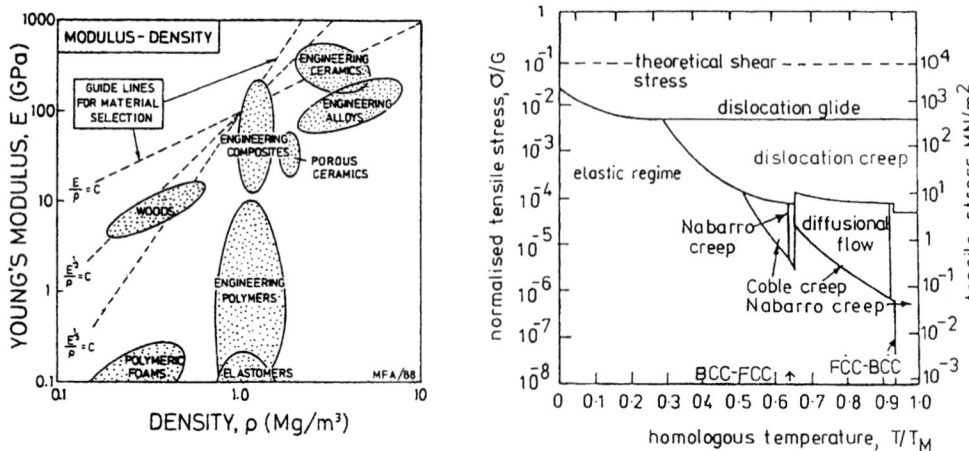

Fig. 6   A materials selection map, Young's modulus of elasticity versus density, is shown in (a) the mechanism map for α-iron (grain size 32 µm) shown in (b) describes the stress and temperature regimes to provide guidance for extrapolation to service conditions for structures and components.

within the design envelope. The potential differences are in loads applied, material/properties, defectiveness, displacements, time of operation, etc. For example, loads are usually controlled in items like pressure vessels, but are significantly more variable for structures such as bridges. Material properties change with age, concretes creep and spall, metals embrittle due to either thermal ageing or neutron irradiation. Structures operating at high temperature are subject to other degradation mechanisms such as creep or oxidation. Degradation can become apparent with regular non-destructive inspection and monitoring of the plant. In order to have confidence in all these situations, it is extremely helpful to develop a good mechanistic understanding. What is more difficult is to point to specific procedures or tools to assist in this process. Mechanistic understanding has been the subject of materials research for many years and many fine books now exist to provide help in this area, e.g. Knott,[20] Hertzberg.[21] For a mechanistic understanding of fracture mechanics assessment, R6 has numerous references to source documents which have provided validation and theoretical underpinning of the main procedures and alternative methods.

PROBABILISTIC METHODS

Finally in this section the use of probabilistic assessments is introduced. The objective of structural integrity assessment is to ensure that the structure or component of interest has sufficient reliability. Reliability implies that a certain probability of failure exists, but this must be minimised to provide sufficient confidence in safe operation. Traditionally, the codes and procedures have concentrated on deterministic methods since this is the least ambiguous way of presenting the 'recipe books'. Unfortunately, nature is not a set of neat compartments with precise boundaries. A 'lower bound' property seldom means precisely that, usually it is a value chosen as being unlikely that lower values would be observed. Analyses, geometries, material properties and loadings are not precise, but all must combine to induce failure. Logically, therefore, the correct assessment of integrity should be by probabilistic methods. Indeed, where the loading is quite clearly of a statistical nature e.g. aircraft loadings, bridge loadings, these methods have found more support. What has restricted the wider application of probabilistic assessment is the high reliability demands placed upon many structural components. This puts great emphasis on the accuracy of tails of assumed distributions if any value is to be placed on the quantified results of such calculations. These are of course the very areas where evidence for the assumed distributions is most lacking. Nevertheless, probabilistic assessment methods are being taken more seriously especially in assisting decision making where relative results of different conditions can be accepted with a great deal more confidence than absolute values.

REQUIREMENTS OF INDUSTRIES

The previous section has provided an overview of the various tools available for generic structural integrity assessment. This section will try to isolate the specific needs of particular industries to understand how the basic tools have been adapted and adopted.

AEROSPACE

The particular needs of the aerospace industry start with material selection since a high strength-to-weight ratio is of enormous commercial importance to the industry. An aircraft is made up of components having wide varieties of loading spectra. For example, the fuselage panels are subjected to a very different load history from that of the landing gear. What is common to both though is the inability to predict precisely the load magnitude for any particular flight. The elements of structural integrity assurance in the aerospace industry are described in an excellent review paper by Goranson.[4] In this paper, it is stated that structural strength in the absence of deterioration such as corrosion fatigue or accidental damage is founded on evolved design criteria. Design limit loads are based on millions of commercial airplane flights. The capability to meet these requirements is demonstrated by analysis supported by tests.

Experience shows that the undamaged structure has, using this design philosophy, a high level of integrity for design operational loads. However, the industry, stimulated by a number of high profile disasters especially of the Comet airliners, realised that service damage was highly significant. In consequence, the Federal Aviation Requirements (FARs) provide detailed requirements for structural integrity assurance. The FARs require that (i) the expected life of components and structure is demonstrated to be greater than the design life of the aircraft, (ii) components and structures should be damage tolerant, (iii) structures should be fail-safe and (iv) where structures experience fatigue damage, the residual strength should always be greater than the maximum expected loads subject to some additional safety factors. The components are supported by a highly structured monitoring and inspection programme.

GROUND TRANSPORTATION

There are two major areas of ground transportation, rail and road. The application of generic tools described in the section, 'Tools Available' has, however, resulted in the statistical fact that only a very small percentage of reported failures in most transport industries is caused by mechanical failure.[22] Especially within the motor industry which produces thousands of nominally identical models, one of the most convincing ways of ensuring structural integrity is through prototype component and assembly testing with loading regimes more demanding than expected service loads. Testing is nevertheless expensive especially as a development tool and finite element analysis is used more frequently at the design stage. Material selection plays a great part in the competitiveness of products but material changes do produce their own structural integrity demands.

A recent review of mechanical failures and structural integrity issues in transport was produced by Smith.[23] It was argued that the considerable expenditure of resource in designing against fatigue had been largely successful in combating the problem. That effort was a combination of experience, conservative design codes, empirical understanding and the easy availability of stress analysis tools. Smith also suggested that remaining weak links are a lack of real service data of the stresses experienced by many components and, in-service, the lack of sensitivity and reliability of crack detection and measuring devices needed to manage inspection periods and to retire components from service. This is clearly of significance with respect to the recent rail failures at Hatfield.

CIVIL STRUCTURES

Civil structures are generally passive but their diversity is enormous. Material selection, as ever, plays an important part of the ultimate assurance of integrity. Seismic tolerance depends on energy-absorbent or flexible materials such as steel or wood rather than stone or bricks. The development of high rise buildings was possible with the advent of high strength-to-weight ratio materials. Concrete allowed more fanciful aesthetic design of buildings and bridges to emerge. However, whichever material is chosen, the same combinations of experience, design codes and advanced analysis are employed. The loading spectra of civil structures are not precise and the statistical nature of some of the more demanding ones such as seismic, wind and wave loading has promoted a move towards probability assessment of some of the major civil structures e.g. within EuroCode for bridges. In most cases it is possible to monitor the loading and response of structures, but this is not general practice unless some particularly critical consequence or weakness is suspected.

NUCLEAR

There are some structures and components within the nuclear industry, failure of which would have far-reaching consequences. In those, however, the loading is generally very closely controlled, material properties are monitored and in most cases detailed routine non-destructive inspection is applied. Nevertheless, access to repair or replace components is often extremely limited. It is this combination of conditions that led to the development of the R6 defect assessment procedure.

The nuclear industry as a consequence concentrates more on sophisticated analytical methods for underwriting structural integrity than any other. In addition to the access difficulties referred to above, the critical structures are usually very large and expensive to produce so that full scale testing is not a feasible avenue for supporting structural integrity. Flewitt[2] described the methods used in one part of the nuclear industry for providing assurance of structural integrity.

PETROCHEMICAL

The petrochemical industry is characterised by having well-controlled loading conditions but of a very wide variety. The various components must be designed, not only from a structural integrity viewpoint, but also with a high regard for the content they are expected to retain. Thus material selection is largely dominated by chemical rather than structural consideration. The industry has ensured that appropriate design codes have evolved for initial design, but the service history may lead to a number of excursions outside the design envelope if not properly monitored. Examples of material property degradation are thinning through corrosion, undetected defects, creep damage and so on. Hence there is a requirement for regular inspection of petrochemical vessels. Repair and/or replacement of petrochemical vessels is more feasible than in the nuclear industry but it is nevertheless potentially expensive and a single vessel downtime can lead to the loss of a complete systems production. Thus the risk based inspection philosophy has developed to provide a balanced view of the places where inspection should be focused to ensure safety at maximum efficiency commensurate with the risk, see for example Francis et al.[24]

DEVELOPMENTS

RISK-BASED INSPECTION

A range of techniques and procedures is adopted to enable a framework for structural integrity assessments of structures and components to be undertaken. As described in the two previous sections these are both wide-ranging and, indeed, tailored to the requirements of specific industry sectors. However, there have been developments in recent years that are directed to improving the confidence in the assessments, reducing the conservatisms to levels appropriate to the overall integrity required and minimising the risk. In the latter context it has to be remembered that risk, R, is

$$R = (L)(C) \tag{1}$$

where L is the likelihood of failure and C the consequence. Hence the likelihood is usually evaluated as a probability, whereas the consequences can be set in the context of a cost or other parameters. Overall, this forms part of the overall balance between economic life, basic cost and overall safety. In the area of structural integrity assessment, this has led to methodology developments in the two main areas for inspection and assessment.[25]

Risk-based inspection methodologies have been developed because non-destructive and other inspections are undertaken to confirm the quality of manufacture and fabrication of structures and components prior to entering service. However, once in service, inspections are undertaken against a declared strategy to detect the onset of any service-induced degradation or potential failure. The approach, although to date generally qualitative, allows a risk matrix to be established, Fig. 7, that provides an evaluation of the need to inspect against a five point test. The approach offers an opportunity to focus inspection on those items where it is most effective and eliminate unnecessary and ineffective inspection. Finally, it provides a systematic and transparent approach for assessors and regulators, manufacturers and, in particular, operators of the plant with the potential to optimise both inspection intervals and plant life.

CRACK ARREST

Most structural integrity assessments assume the possible presence of defects, taken to be cracks, in the structure or component. As described above, fracture

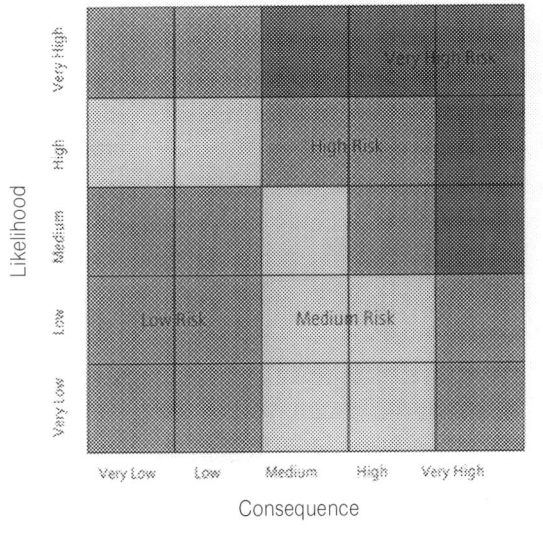

Five Point Test

1 Was the manufacturing and fabrication inspection adequate

2 Is the component subject to in-service degradation?
   If yes:

3 Will the identified degradation mechanism(s) lead to failure?

4 Is a valid inspection or monitoring procedure available?

5 Does it matter if the component fails?

Risk = Likelihood x Consequence

Fig. 7 Risk-based inspection ranking including the five point test.

mechanics is used to identify conditions where defects could lead to failure of the structure or component usually defined by crack initiation.[26] Figure 8 shows a simple flow diagram that describes the interrelationship between this initiation based approach and an alternative crack arrest approach. The crack arrest concept recognises that not every crack initiation event leads inevitably to failure of the structure or component, even when the fracture mode is brittle. Indeed there is a logical relationship between these two approaches, where arrest could be adopted as an alternative approach or the two could be combined into a single integrated assessment as shown by Fig. 8.

The structural integrity assessment of Magnox reactor steel pressure vessels described elsewhere[27] seeks to show that conditions under which fracture could initiate in the vessel do not exist, i.e. an initiation-based approach. This contributes to the overall safety case for the vessels which is as based upon a multi-legged argument set in a defence-in-depth framework described by TAGSI.[28] For the assessment of reference defects, fracture mechanics is applied to postulated defects chosen because the likelihood of any real defects of greater severity remaining undetected following construction is extremely small. Here the procedure adopted is R6 where the main inputs to the assessment are (i) depth, shape and position of a potential defect, (ii) materials properties, (iii) loading conditions and (iv) geometry of the component. Reference defects 25 mm deep and fully extended are invoked, except for special cases where a 6:1 aspect ratio is

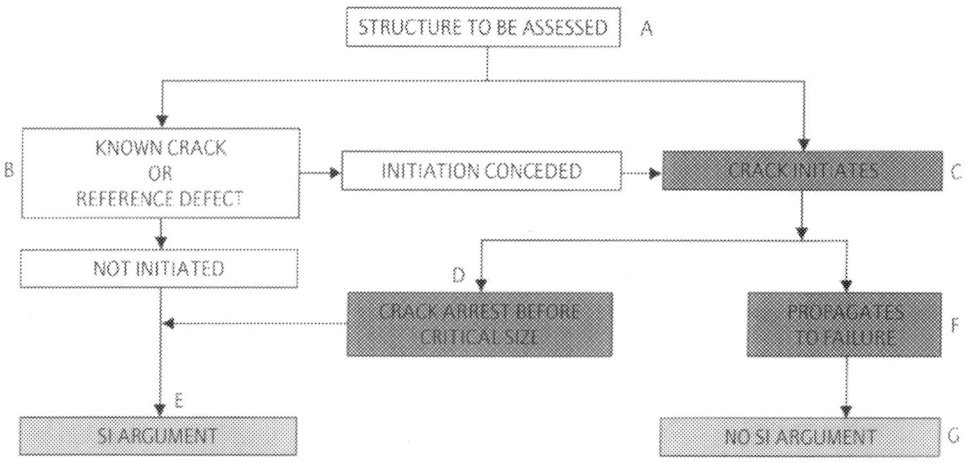

Fig. 8 Relationship between initiation and crack arrest structural integrity assessment methodologies.

adopted. The calculations are carried out using statistical lower bound values of fracture toughness and tensile properties estimated from a data set. The assessment additionally focuses on the ductile to brittle transition temperature and the approach is to operate the vessels at a temperature on the shelf of this transition curve; a ductile regime. The assessment procedure is used to determine a minimum failure pressure as a function of vessel temperature defined by a pressure limit line, for different locations on the vessel, Fig. 9. By comparing this failure avoidance line with the pressure/temperature limits set by the Operating Rule, it is possible to (i) ensure that at normal operation the vessel is at a temperature at or above the onset of the upper shelf temperature of the ductile-to-brittle transition curve and (ii) adequate reserve factors on pressure are achieved so that crack initiation will not occur. These structural integrity assessments are undertaken at regular intervals over the operating life to accommodate, for example, changes of mechanical properties due to exposure to the service environment of neutron irradiation and temperature, $\leq 360°C$.

Connors et al.[26] have reviewed a number of observations of crack arrest in various structures described in the literature. Table 1 highlights typical conditions under which crack arrest can be expected. In case (A), C–Mn steel pressure vessels were pressurised to different levels, and the material adjacent to the defect cooled with liquid nitrogen. Pressurisation resulted in ligament failure with some axial propagation of the fully penetrating defect. However, when the temperature of the

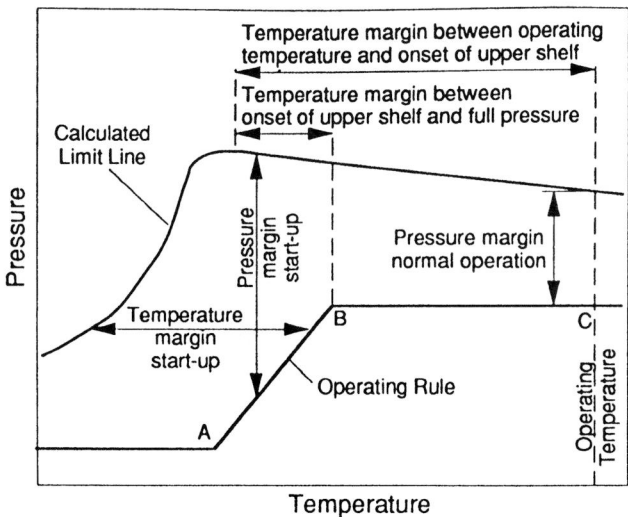

Fig. 9 A schematic diagram showing the pressure and temperature margins that can be derived when the Rule to which the plant is operated is compared with the calculated limit line.

Table 1   Crack Arrest Observations

Item	Observed Crack Arrest	Ref.
(A)	Ligament failure of long axial partly penetrating defects in cylindrical pressure vessels	29
(B)	Propagation of a fully-penetrating defect in cylindrical pressure vessels which had been strain aged	30
(C)	Local failure of a low toughness weld	31
(D)	Propagation of a crack away from a nozzle	32
(E)	Pop-in crack propagation from a fully-penetrating axial defect in a cylindrical pressure vessel	33
(F)	Radial propagation of a long axial partly penetrating defect during a thermal shock test on a thick cylindrical pressure vessel	34
(G)	Axial propagation from a semi-elliptical defect in a spinning cylinder during a thermal shock test	35, 36

surrounding material was above a certain value related to the level of applied stress, cracks were observed to arrest. The vessels in case (B) were subject to a temperature and load history designed to cause strain ageing embrittlement at tips of fully penetrating axial defects. Rapid propagation of cracks occurred under pressure, but some arrested when the tips had passed through the embrittled region into tougher steel beyond. In case (C), the brittle failure of a weldment occurred during pressurisation of a thick-walled cylindrical vessel containing an internal axial, partly penetrating defect which intersected the weld. The crack arrested before the wall was breached; the pressure vessel steel on either side of the weldment was of sufficient toughness to enable the small crack emanating from the weldment to cease propagating. In case (D) the propagation and arrest of a defect from a nozzle was the result of a crack growing into a region of reduced stress. Cases (F) and (G) refer to cracks subject to thermal shock conditions; where propagation from the colder, and more highly stressed, layers of the vessel wall can be arrested when warmer, tougher material is reached. Item (G) further indicates that an important factor for crack arrest is the shape adopted by the defect, which underwent rapid axial propagation without deepening. This is also illustrated by the crack arrest profile of the fully penetrating defects in case (E), which became highly curved. Similar curved crack fronts have been observed in arrested defects in large plates loaded under both quasi-static and impact conditions.[37, 38]

In summary, an important factor in the arrest of cracks in cases (A), (B), (C) and (F) is the existence of a region of low toughness at the crack tip, surrounded by tougher material. Defect shape changes are significant in (D), (E) and (G), and are a contributory factor in the thermal shock examples (F) and (G). Thus applications of crack arrest concepts in structural integrity arguments are likely to involve locally brittle regions embedded in tougher material, and defect configurations giving rise to reductions in stress intensity factors as crack extension occurs. Connors et al.[26] concluded that a promising specific application of a crack arrest structural integrity based argument would be for a cold thermal shock load applied to a PWR pressure vessel during a loss-of-coolant accident. This combines a reducing stress intensity factor and an increasing fracture toughness, both of which favour arrest. In Magnox reactor steel pressure vessels described above, thermal shocks are much less severe, so the potential for significant and favourable fracture toughness and stress intensity factor gradients is lacking. Fracture toughness gradients can arise, however, as result of through-the-wall variations in neutron radiation damage, or a postulated local microstructural region of brittle material. In these examples, crack arrest may potentially be an ingredient in a structural integrity assessment, but this would require evaluation through a consideration of the spatial distribution of fracture toughness. Again, in this particular case, the amount of additional data and information required to produce a full crack arrest safety argument could be considered prohibitive but, for supporting arguments, there was the potential for obtaining adequate information in a timescale commensurate with the need. Applications based on the possible reduction of stress intensity factor as a result of crack shape changes may be considered too specialised to be operational at this stage. Shape changes are frequently observed experimentally, but at present there is no reliable method for predicting them. Short duration loads have potential for analysis on the basis of crack arrest, but the duration must be sufficiently short, and the dynamic response of the structure must be evaluated.

MECHANISTIC CONSIDERATIONS

In general, deterministic arguments for high integrity structures or components adopt conservatisms that could be relieved if there was a better understanding of potential defects together with materials properties and the changes over the service life of plant subject to time dependent degradation processes. Certainly improved mechanistic understanding would allow increased confidence in the prediction of operating life of structures and components. In general, there is a need to improve the ability to predict the performance of a material from a knowledge of composition, microstructure, processing and operational history. Although surveillance specimens assist assessments and facilitate underwriting the integrity of structures and components, there remains a need to improve both the formulation and interpretation of accelerated life testing, and the associated methodology. This means that accumulated life test data for a range of temperatures, environments and stress states have to be combined with predictive assessment methodologies developed beyond the present understanding. For example, when assessing high integrity structures and components it is not sufficient to simply adopt materials behaviour derived from accelerated and, therefore, artificial test conditions. There is a need to develop structural integrity assessment methodologies and input data which, when combined with a mechanistic model, will allow quantitative assessment of life with an appropriate level of confidence to ensure failure will not occur, and provide the necessary level of reliability. Until this is achieved, it is necessary to incorporate the high cost conservatisms into the assessment methods.

PROBABILISTIC METHODS

An alternative is to consider probabilistic structural integrity assessments either in total or as part input to a deterministic argument. Detailed probabilistic fracture mechanics methodologies based upon, for example, the R6 procedure are being developed,[39-41] Fig. 10. In general these require a probability density function for each parameter and take the simple form

$$Pf = \int_{o}^{W} f(a) \iint_{A\,fail} P_1(K_{Ic})P_2(\bar{\sigma})\mathrm{d}\bar{\sigma}\mathrm{d}K_{Ic}\mathrm{d}a \qquad (2)$$

Where $f(a)$, $P(K_{Ic})$ and $P_2(\bar{\sigma})$ are the probability density functions for a defect of size a, the fracture toughness, K_{Ic}, the flow stress, $\bar{\sigma}$, of the material. The limits of integration are in relation to the section thickness and the failure region of the R6

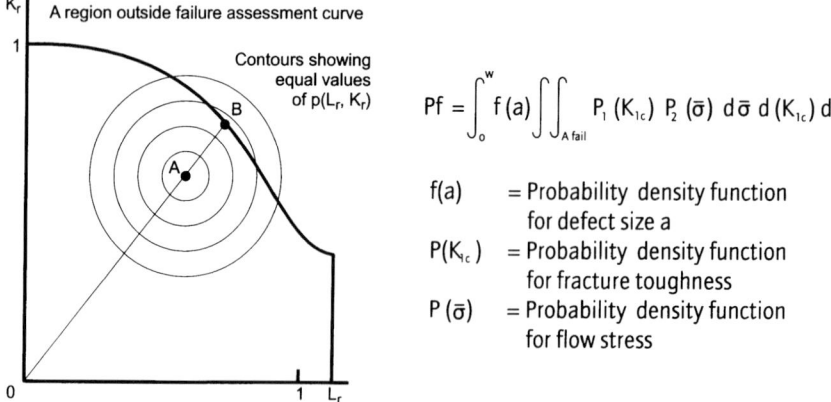

Fig. 10 A probabilistic R6 failure avoidance diagram. The integration limits are section thickness and the failure region of the R6 diagram and then reserve factor on load is $F^L = OB/OA$.

diagram. In practice, even at this relatively simple level, there is a requirement to have knowledge of the defect distribution obtained from non-destructive examinations. In this respect Wilson[41] has been developing probabilistic considerations of such defect distributions based upon inspection results combined with the confidence in the measurement technique. For example, to apply this probabilistic methodology in full to the steel reactor pressure vessel assessments, it would be necessary to take account of factors such as, critical crack length, pre-service proof test, fault loadings, stable tearing, fatigue crack growth and competing ductile and brittle fracture mechanisms.

The incorporation of these into the probabilistic argument would add considerably to the complexity of the expression to be evaluated and the probabilistically distributed data would have to be acquired, but the basic principles remain. Various statistically based procedures have been and are being adopted to provide appropriate materials property data for such methodologies. For example Moskovic and Flewitt[42] used non linear regression in a Bayes framework to analyse Charpy impact energy data obtained for C–Mn steel weldments, where the product is a set of probability distributions for the impact energy curves and the shift in the transition temperature. In addition to this approach, neural network techniques are being used to provide materials mechanical property data. A neural network is a general method of regression analysis in which a very flexible, non-linear function is fitted to experimental data.[43, 44] In this way, it is potentially possible to use the neural networks to make predictions and establish how these depend upon various contributions of input data. In this respect, Lalam et al.[45] have considered the mechanical properties of C–Mn ferritic steel weldments and found it possible to create reasonable neutral network models for yield, UTS, elongation and Charpy impact energy properties, However, there are still considerable developments needed in this area before the true potential of this methodology will be obtained. At present it is more realistic that these methods should be considered and used to enhance the traditional deterministic structural assessments.

CONCLUSIONS

(i) Over the past 30 years there have been very significant developments in the tools and methodologies for analysis and the mechanistic understandings of structural behaviour.

(ii) These developments allow designers and operators to work closer to limits. Hence there is a need to have the necessary confidence in input data, recognition of the capabilities of the analysis tools and a responsibility to operate within the design envelope.

(iii) The definition and understanding of input parameter distributions needs to be improved for probabilistic methods to achieve their promise in improving the assessment of structural reliability.

ACKNOWLEDGEMENTS

The authors would like to thank their various colleagues who have contributed to the views expressed in this paper. It is published with the permission of the Head of Reactor Services BNFL Magnox Generation.

REFERENCES

1. G. Galileo, *Discorsia Dimonstrazioni Matematiche*, Leiden, 1638.
2. P. E. J. Flewitt, *Structural Integrity Assessment of High Integrity Structures and Components: User Experience, Mechanical Behaviour of Materials*, A. Bakker ed., Delft Univ. Press, (Delft), 1995.
3. J. F. Knott, *Two Steps from Disaster – The Science and Engineering of Structural Integrity*, The Royal Society and The Royal Academy of Engineering 1999 Lecture, Royal Soc. and Royal Acad. Eng. (London), 1999.
4. U. G. Goranson, *Fatigue*, 1994, **16**, p. 43.
5. D. O. Harris, C. H. Wells, S. A. Rau and D. D. Dedhia, *Int. J. Pressure Vessels and Piping*, 1994, **59**, p. 175.
6. B. Neubauer and F. Bietenbeck, *Int. J. Pressure Vessels and Piping*, 1989, **39**, p. 57.
7. T. Endo, *Int. J. Pressure Vessels and Piping*, 1994, **57**, p. 7.
8. T. Swift, *Fatigue*, 1994, **16**, p. 75.
9. J. Hodgson and C. M. Boyd, 'Brittle Fracture in Welded Ships – an empirical approach from recent experience, *Quarterly Trans. INA*, 1958, **100** (3).
10. J. Lancaster, *Engineering Catastrophes*, Abington Pub., 1994.
11. D. Kalderon, *Proc. Inst. Mech. Eng.*, 1972, **186**, p. 341.
12. T. Grey, *ibid.*, 1972, **186**, p. 339.
13. J. Wannenberg, G. C. Klintworth and A. D. Rauth, *Int. J. Pressure Vessels and Piping*, 1992, **50**, p. 255.
14. Warren C. Young, *Roark's Formulas for Stress and Strain*, McGraw Hill, 1986.
15. *Guidelines to Finite Element Practice*, NAFEMS, Department of Trade and Industry, 1984.
16. *R6 'Assessment of the Integrity of Structures containing Defects'*, Revision 4, British Energy Generation Ltd, 2001.
17. D. P. Brown, 'Observations on experience with welded ships', *Welding Journal*, September 1952.
18. R. P. Harrison, K. Loosemore and I. Milne, *Assessment of the Integrity of Structure containing Defects*, CEGB Report R/H/R6, 1976.

19. M. F. Ashby and D. R. H. Jones, *Engineering Materials 2*, Pergamon, 1986.

20. J. F. Knott, *Fundamentals of Fracture Mechanics*, Butterworth, 1973.

21. R. W. Hertzberg, *Deformation and Fracture Mechanics of Engineering Materials*, Wiley, 1989.

22. S. Mathews, 'Human Factors in Aviation Safety', *Proc. Int. Workshop on Technical Elements for Aviation Safety*, H. Jereda ed., Nat. Aerospace Lab, Tokyo, 1999.

23. R. A. Smith, 'Transport Safety. The Role of Machines and Men', *Proc. 5th Int. Conf. on Engineering Structural Integrity Assessment*, EMAS, 2000.

24. A. Francis, R. J. Espiner, A. M. Edwards and R. J. Hay, 'A consideration of Data Requirements for Structural Reliability based Assessments of Onshore Pipelines', *Proc. 5th Int. Conf. on Engineering Structural Integrity Assessment*, EMAS, 2000.

25. B. Tomkins, 'A Historical Perspective of Materials Related Structural Integrity Issues in the Nuclear Industry', *Fracture, Plastic Flow and Structural Integrity*, P. B. Hirsch and D. Lidbury eds, Institute of Materials, 2000.

26. D. C. Connors, A. R. Dowling and P. E. J. Flewitt, *Crack Arrest Concepts for Failure Prevention and Life Extension*, Abington Publishing, 1996, Paper 13.

27. P. E. J. Flewitt, G. H. Williams and W. J. Graham, *The Inspection and Validation of Nuclear Power Plant*, I. Mech. Eng., 1994, p. 11.

28. *Incredibility of Failure Safety Cases*, R. Bullough ed., TAGSI Report TAGSI/P(97) 140 Rev. 6.

29. B. Edmondson, C. L. Formby, R. Jurevics and M. S. Stagg, *Aspects of the Failure of Large Steel Pressure Vessels*, CEGB Report RD/B/N1295, 1969.

30. C. L. Formby and W. Charnock, *The Effect of Strain Ageing Embrittlement Upon Pressure Vessel Integrity*, CEGB Report RD/B/R1846, 1971.

31. R. Rintamaa, H. Keinanen, K. Torronen, H. Talja, A. Saarenheimo and K. Ikonen, 'Fracture Behaviour of Large Scale Pressure Vessels in the Hydro Test', *Int. Journal of Pressure Vessels and Piping*, 1988, **34**, p. 265.

32. J. G. Merkle, G. C. Robinson, P. P. Holz and E. J. Smith, *Test of 6-Inch Thick Pressure Vessels, Series 4; Intermediate Test Vessels V5 and V9 with Inside Nozzle Corner Cracks*, ORNL NUREG Report 7, 1977.

33. R. W. Nichols, W. H. Irvine, A. Quirk and E. Bevitt, UKAEA TRG Report 004 (C).

34. J. Sievers and H. Schulz in Collaboration with B. R. Bass and C. E. Pugh, and J. Keeney, *FALSIRE Phase 1 Comparison Report*, NUREG/CR-5997, 1994.

35. D. J. Lacey, R. E. Leckenby, E. Morland, M. Hamid, K. A. May, I. J. O'Donnell, A. M. Clayton, E. R. Johnson, B. R. Bowdler, J. T. Bland and J. B. Stones, *Spinning Cylinder Test 4: An Investigation of Transition Fracture Behaviour for Surface Breaking Defects in Thick Section Steel Specimens*, AEA Report TRS 4098, 1991.

36. K. A. May, E. Morland, D. J. Lacey, B. R. Bowdler, R. P. Birkett, M. Hamid, A. M. Clayton and J. B. Stones, *Spinning Cylinder Test 6: A Further Investigation of Transition Fracture Behaviour for Surface Breaking Defects in Thick Section Steel Specimens*, AEA Report AEA/RS/4305, 1993.

37. C. S. Wiesner, B. Hayes, S. D. Smith and A. A. Willoughby, 'Investigations into the Mechanics of Crack Arrest in Large Plates of 1.5% Ni TMCP Steel', *Fracture and Fatigue of Engineering Materials and Structures*, 1994, **17**, p. 221.

38. R. W. Nichols, 'Fast and Brittle Fracture Studies Related to Steel Pressure Vessels', *Proc. Roy. Soc. (London)* A, 1965, **285**, p. 104.
39. R. S. Gates, A. Francis, M. Kolbuszewski, R. Wilson and P. L. Windle, Nuclear Electric Report, TD/SID/REP/0030, 1990.
40. R. Wilson and R. A. Ainsworth, 'A Probabilistic Fracture Mechanics Assessment Procedure', ISMIRTH Conference, Tokyo, 1991, paper G50M.
41. R. Wilson, 'The Use of Probabilistic Fracture Mechanics in Support of Deterministic Assessments of Structural Integrity', *Lifetime Management and Evaluation of Plant Structures and Components*, J. H. Edwards ed., EMAS, 1998.
42. R. Moskovic and P. E. J. Flewitt, *Met. Trans.*, 1997, **28A**, p. 2609.
43. D. J. C. Mackay, *Neural Computing*, 1992, **3**, p. 448.
44. H. K. D. H. Bhadeshia, 'Models for Elementary Mechanical Properties of Steel Welds', in *Mathematical Modelling of Weld Phenomena*, H. Cerjak ed., Institute of Materials, 1997, pp. 229–284.
45. S. H. Lalam, H. K. D. H. Bhadeshia and D. J. C. Mackay, 'Estimation of Mechanical Properties of Ferritic Steel Welds, Part 2: Elongation and Charpy Toughness', *Science and Tech. of Welding and Joining*, 2000, **5**, pp. 149–160.

CHAPTER 2

Methods for the Assessment of Structural Integrity of Components and Structures – Regulator's View – Nuclear Applications

A. J. Cadman
Nuclear Installations Inspectorate, Health and Safety Executive, St Peter's House, Balliol Road, Stanley Precinct, Bootle L20 3LZ, UK

ABSTRACT

The Health and Safety Executive (HSE) regulates the nuclear industry in the United Kingdom via its Nuclear Installations Inspectorate (NII). NII oversees a wide range of nuclear sites including Magnox power stations, Advanced Gas Cooled Reactor power stations, a Pressurised Water Reactor and the Sellafield site (over 100 assorted plants). These sites have a wide diversity of structural integrity issues.

HSE's and therefore NII's approach to regulation is one of goal setting rather than prescription. All of the licensed sites have a licence with 36 standard conditions. NII's goal setting intent allows each licensee to develop compliance arrangements which best suit their business so that whilst the conditions are the same for each licence the detailed arrangements for meeting these conditions vary from licensee to licensee.

HSE has produced a set of Safety Assessment Principles (SAPs) which have been published. SAPs are used by NII to establish whether a safety case is adequate. Publishing the SAPs provides transparency to the licensees so that they know what NII expects from them. The SAPs also help to provide a consistent approach from NII's inspectors. The SAPs cover all the relevant topic areas for nuclear regulation including structural integrity. Unlike the approach used by the United States Nuclear Regulatory Commission the SAPs do not specify that certain codes or methods must be used. Instead the SAPs provide targets that need to be met by a nuclear safety case. For structural integrity cases these targets vary depending on the contribution that a structure, system or component makes to nuclear safety. For items such as reactor pressure vessels where the vessel is the principal means of ensuring safety a Special Case procedure is followed.

Although NII does not demand that specific assessment methods are used it still has to be convinced that a licensee has an adequate safety case. NII therefore challenges new safety cases and methodologies as they are presented. The extent of the challenge will depend on the difference in the current case and methodologies compared to those that have been previously assessed.

NII does not work in a vacuum but takes advice and information from numerous sources. The Health and Safety Commission (HSC) has a nuclear safety advisory committee (NuSAC formally ACSNI). This group advises HSC on major issues affecting the safety of nuclear installations, for example NuSAC has previously provided NII, through HSC, with advice on the suitability of the R6-Rev 3 methodology for use in assessing the integrity of structures containing defects. NII is also one of the sponsors of TAGSI (the UK Technical Advisory Group on the Structural Integrity of nuclear plant). It has asked TAGSI a range of questions associated with structural integrity. NII also gathers information from research. In the United Kingdom there is a joint programme of nuclear safety research agreed between NII and the major members of the nuclear industry. Some of this research is directed towards developing methodologies for future safety cases and evaluating conservatism in current assessment methodologies.

INTRODUCTION

The Health and Safety Executive (HSE), via its Nuclear Installations Inspectorate (NII), regulates the nuclear industry in the UK. NII was originally established in 1960 to regulate the then growing civil nuclear industry. It became part of HSE in 1974 when the various separate inspectorates that regulated safety in different industries came under the overall control of HSE. HSE obtains its primary powers from the Health and Safety at Work etc Act 1974.[1] Under this employers have duties to protect both their employees and other people from their work activities.

The Nuclear Installations Act 1965 (as amended)[2] is a relevant statutory provision of the HSW Act. This provides the HSE with its powers to license the use of a site for the installation or operation of a nuclear installation. The Chief Inspector of Nuclear Installations, on HSE's behalf, has the power to attach conditions to site licences as appear necessary or desirable in the interests of safety or as it may think fit with respect to the handling, treatment and disposal of nuclear matter.

All nuclear site licences have the same 36 standard conditions attached to them. These conditions concern every aspect of nuclear safety at the plant including the construction or installation of new plant, modifications to existing plant, operating limits and safety case production. The site licence conditions provide NII with a wide range of powers including the powers to grant or issue (a) Consents which are required before any activity specified in the licence is carried out, (b) Approvals for any arrangements under the licence that NII decides requires its approval (c) Directions to legally require the licensee to take a particular action such as shutting down a plant.

These licence conditions provide a strong regulatory oversight of the nuclear industry through a non-prescriptive permissioning regime and NII is therefore a non-prescriptive regulatory organisation. UK health and safety law is generally goal setting – setting out what must be achieved, but not how it must be done. On a wide range of health and safety issues HSE provides codes and guidance on good practice. Neither HSE's codes nor its guidance material are in terms that will necessarily fit every case. In considering whether the legal requirements have been met, Inspectors take the relevant codes and guidance into account, using sensible judgement about the extent of the risks and the effort that has been applied to counter them. This non-prescriptive approach is in contrast to the one taken in the United States where in the nuclear field the US Nuclear Regulatory Commission (USNRC) prescribes the required standards.

For example the US Code of Federal Regulations[3] prescribes that the systems and components of boiling and pressurised water-cooled nuclear power reactors must meet the requirements of the appropriate parts of the ASME Boiler and Pressure Vessel Code. Thus at the time of writing it is the 1995 Edition through to the 1996 Addenda that are the relevant parts of Section III of the ASME Code (subject to numerous limitations by USNRC) that nuclear operators have to comply with. The relevant edition of ASME required to meet the USNRC requirements is updated periodically in the Code of Federal Regulations.

SAFETY ASSESSMENT PRINCIPLES (SAPs)

NII regulates a broad range of nuclear sites including Magnox power stations, Advanced Gas Cooled Reactor power stations, the Sizewell B Pressurised Water Reactor (PWR) and the Sellafield site (over 100 assorted plants). These sites have a wide diversity of structural integrity issues.

The process adopted by the NII in making permissioning decisions requires a safety case and supporting reports to be submitted by the licensee for assessment by NII. The inspectorate needs to adopt a consistent and uniform approach to the assessment process; to this end it is necessary to provide a framework that can be used as a reference for the technical judgements that the assessors have to make. HSE's 'Safety Assessment Principles for Nuclear Plants' (SAPs) provide such a framework.[4] In some subject areas the SAPs embody specific statutory limits. Apart from these, NII assessors are expected to apply judgement, based on their knowledge and experience, in the interpretation of the SAPs. There is therefore no rigid interpretation of the principles. However, the engineering principles in particular represent the NII's views on good modern practice and NII would not expect modern plants to have difficulty in satisfying them. For the assessment of 'old plants' there is a further point to be considered in that the safety standards used in their design and construction may differ from those used in more modern plants such as the Sizewell B PWR. The existence of such differences has to be recognised by NII's assessors when applying the SAPs in the assessment of older plants. Licensees need to demonstrate that the risk from the plant has been reduced 'as low as is reasonably practicable' (ALARP). The ALARP principle is of particular importance when assessing older plants, when the age of the plant and its projected life become important factors to be taken into account when making judgements on the reasonable practicality of making improvements to those plants. Again this can be contrasted with the USNRC approach where the specification for the use of a particular ASME code contains a large number of limitations and modifications to the exact parts and ages of the code, which are applicable depending on the age of the reactor plant.

SAFETY CATEGORISATION

The starting point for the assessment of a structural integrity component is its safety categorisation (SAPs 69 and 83). The categorisation takes account of the consequences of failure and the failure frequency requirements. All structures important to safety need to be designed, constructed and inspected to the best practicable standards commensurate with their safety categorisation.

Category 1 is for any structure, system or component that forms a principal means of ensuring nuclear safety. This includes major components such as reactor pressure vessels. For Category 1 components the SAPs require that conservative design and construction standards should be adopted for this, the highest category together with a strict interpretation of the SAPs in line with the ALARP requirement. Where a structure, system or component forms a principal means of ensuring nuclear safety and it is not practicable to demonstrate that the accident frequency principles are satisfied in the event of its failure, the plant may only be accepted after the application of a special case procedure agreed as an alternative. The special case procedure for structural integrity components is discussed later.

Category 2 is for any structure, system or component that makes a significant contribution to nuclear safety. For these components appropriate national or international codes or standards are expected to be adopted, with particular consideration being given to demonstrating the ability of the item to perform the required safety function.

Category 3 structures are any other structure, system or component and normal industrial standards can be applied.

REQUIREMENTS FOR ASSESSMENT METHODS

Where there is no appropriate code or standard NII expects a full justification for the design method adopted. The combining of different design codes and standards for an individual component has to be avoided where practicable and needs to be justified when used.

Whatever category is required due allowance has to be made in the design for degradation processes, including corrosion, erosion, creep, fatigue, and ageing, and for the effects of the chemical and physical environment. The design has to allow for any uncertainties in determining the initial state of components and the rate of degradation.

The SAPs have certain expectations for theoretical and analytical models and methods. Where stress analysis is used to support the design it needs to demonstrate that the component has an adequate life, taking into account time-dependent processes. The analysis needs to use methods that have been verified and validated, using model tests if necessary. The data used in any analysis need to be demonstrably conservative. In particular, the uncertainties associated with material properties affected by degradation should be taken into account. Where appropriate, studies need to be carried out to determine the sensitivity of analytical results to the assumptions made, the data used, and the methods of calculation.

For components subject to the special case procedure, and where appropriate for other metal components, the size of defects of structural concern need to be calculated, using verified and validated fracture mechanics methods. The methods need to be based on a sound scientific understanding and any necessary assumptions or approximations need to bias the results demonstrably in a safe direction. The structural integrity assessment methods need to be validated as a whole or, where this is not practicable, on a module basis, against experiments. Where appropriate, an independent check of an analysis is expected to be carried out using a different method or analytical model.

The data used in the design and fault analysis of safety-related aspects of plant performance need to be shown to be valid for the circumstances by reference to established physical data (e.g. measured materials properties), experiment or other appropriate means. Where uncertainty in the data exists, an appropriate margin in a safe direction should be provided to take account of it. Extrapolation from available data cannot be used without good justification. Licensees need to keep under review both the physical property data and the scientific knowledge behind the methodology to ensure that existing safety cases are not invalidated.

THE SPECIAL CASE PROCEDURE FOR STRUCTURAL INTEGRITY COMPONENTS

The general lack of adequate reliability data for structural components leads to the assessment of this type of component being based primarily on established engineering practice. Even when there is some confidence in assessing the reliability of a structural integrity component on the basis of existing data and a probabilistic safety case is possible, it is unlikely that a case will be acceptable to NII without substantial support from theoretical analyses and engineering judgement. As a result, although the radiological consequences of failure of structural components may be significant, it is often not possible to calculate the risk of failure for inclusion in the Probabilistic Safety Assessment for a plant.

For structural integrity components which form a principal means of ensuring safety, for example a reactor pressure vessel, there are two particularly important aspects that need to be addressed in a safety case:

(i)   the structure should be as defect free as possible; and

(ii) it should be demonstrated that the structure is defect tolerant, in particular the critical crack sizes should be large with respect to the inspection technique.

In order to achieve these fundamental requirements, several related but independent arguments need to be used, based on the following concepts:

(a) the use of sound design concepts and proven design features;

(b) the analysis of the potential failure modes for all conditions arising from design basis faults;

(c) the use of proven materials;

(d) the application of high standards of manufacture, including in-process inspection, and construction, for the materials and processes used;

(e) high standards of quality assurance throughout all stages of design, procurement, manufacture, construction and operation;

(f) pre-service and in-service inspection to detect defects at sizes below those which have the potential for causing or developing into a failure mode, and to size these defects conservatively;

(g) the provision of in-service plant and materials monitoring; and

(h) the existence of a leak-before-break case.

For components, which are not of major safety significance, this list of requirements is also relevant, though the stringency of their application needs to reflect the safety categorisation of the item.

WHAT DOES NII DO TO CONVINCE ITSELF THAT VERIFIED AND VALIDATED METHODS ARE USED BY LICENSEES

CHALLENGING LICENSEES

As part of its normal business NII will examine a licensee's safety case and the methodologies behind its production. NII asks the licensee to provide justifications for the methods that are used. This can require the licensee to do extra work to provide a more comprehensive justification to the one that is already available. In order to assist in its decision making process NII gathers advice and information from many sources. Some of these sources are considered below.

DIRECT CHALLENGE TO THE LICENSEE

As has been described above NII does not prescribe specific codes or assessment methodologies to be used in the production of safety cases. However, it does seek to ensure that certain principles have been met. NII ensures licensees meet these principles through a process of constant challenge as new safety cases are presented. The extent of the challenge will depend on the difference between the new case and previously assessed cases. A novel method in a safety case would attract an in-depth challenge of the validation of the method over many meetings between the NII and the licensee. A safety case, which used recognised and well-understood methodology, would be less likely to have the methodology challenged but there would still be careful consideration of issues such as the input data and the applicability of the methodology to the case. For a methodology such as R6, the use of appropriately pessimistic input data is important. The method will always give an answer and it is the pessimism in the input data that provides the margins on failure that the safety cases rely on.

NUCLEAR SUPPORT CONTRACTS

At times NII considers it appropriate to fund specific contracts for technical advice from outside organisations. For example, NII has previously funded a review of parts of the R5 high temperature assessment procedure to get confidence in a licensee's claims for its use in safety cases.

NuSAC

The Health and Safety Commission (HSC) is responsible to several ministers for the administration of the HSW Act. The Nuclear Safety Advisory Committee (NuSAC, formally ACSNI) advises HSC on major issues that affect the safety of nuclear installations.

When the CEGB launched the third revision to its R6 procedure (R6 Rev 3) ACSNI set up a study group to examine and comment on the CEGB's R6 methodology. Over a period of 8 months the group held a series of meetings. In the report on its findings it considered a range of issues:

- The development of the R6 method to form R6 Rev 3.
- Input data requirements.

- Assessment procedures using R6 Rev 3, sensitivity studies and reserve (safety) factors.
- Validation.
- Comparison with other assessment methodologies.
- Limitations and future developments.

The study group concluded that the R6 Rev 3 methodology represented an acceptable and safe method for assessing the structural integrity of nuclear components within its defined scope. ACSNI subsequently published its opinion and recommendations for components which operated at temperatures below the creep range.[5]

NII has subsequently considered these recommendations along with its own views formed from questioning the R6 procedure as it has been further developed.

TAGSI (THE UK TECHNICAL ADVISORY GROUP ON THE STRUCTURAL INTEGRITY OF
NUCLEAR PLANT)

NII became sponsors of TAGSI in April 1994. We are interested in hearing TAGSI's views on generic structural integrity issues while maintaining our regulatory independence with regard to specific safety cases. Over the last 7 years NII has asked for advice from TAGSI on several generic questions relating to structural integrity:

- issues concerning Incredibility of Failure safety cases (complete);
- the benefits and reliability of warm pre-stressing (complete);
- the effect of neutron irradiation on the ductile tearing toughness of ferritic steels (complete);
- crack arrest: changes in crack length with ligament 'snap-through' (complete);
- the strengths and weaknesses of proof pressure test arguments in structural integrity assessments (ongoing).

The results from the completed work have been used to inform NII on these generic issues. The results have not been used as a substitute for NII's own opinion when coming to regulatory decisions but provide an important input.

RESEARCH

Each major nuclear generating licensee has responsibility to undertake programmes of research to address the safety issues arising from the operation of their nuclear power reactors. In addition, the UK Government has historically commissioned programmes of research to address safety concerns of a generic nature. In 1990, the Department of Trade and Industry transferred its responsibility for commissioning this work to the HSC. HSE manages the *HSC Co-ordinated Programme of Nuclear Safety Research* on HSC's behalf. The arrangements for implementing the HSC Programme require both HSE and the major nuclear generating licensees to commission research programmes to address safety issues identified by the NII in its Nuclear Research Index (NRI).[6] The NRI, which is produced annually, is a compilation of generic nuclear safety issues generated by the NII as a result of its knowledge gained in regulating nuclear reactor sites and its broader dealings with other organisations, both nationally and internationally. Discussions then take place with the major licensees' Industry Management Committee (IMC, made up from representatives of British Energy Generation and British Nuclear Fuels plc) in order that these issues are prioritised in the context of the overall programme strategy. The issues in the NRI associated with methods for assessing structural integrity currently fall into 13 areas:

- treatment of the potential for pipe breaks;
- failure assessment diagram methods for thermal stress;
- improved understanding of proof pressure test argument;
- improved fracture mechanics methodologies;
- improved guidance on stress classification for primary and secondary stress;
- improvement of high temperature creep crack growth estimation methods;
- development of crack arrest models;
- analysis validation;
- development of probabilistic fracture mechanics;
- improved dynamic analysis;
- warm pre-stressing;
- seismic integrity of pipework;
- improved understanding of leak before break arguments.

Not all the issues remain open today as work in some areas is complete. The amount of work done under each issue varies depending on NII's requirements. For example under the issue of 'Improved Fracture Mechanics Methodologies' the following sub-issues have been identified:

- evaluation of the Local Approach method;
- UK/French collaboration in evaluation of the Local Approach;
- Local Approach simulation of two large-scale fracture experiments;
- evaluation of two criteria approaches;
- participation in Phase II of Project FALSIRE (Fracture Analysis of Large Scale International Reference Experiments);

- examination of limitations of increasing load limits by invoking the constraint differences between test specimens and real structures;
- analysis of defects with realistic profiles taking account of local constraint;
- evaluation of method of real elements for integrity assessments;
- linear summation of fatigue and tearing crack growth in austenitic stainless steel;
- ductile tearing methodologies.

The results of research work under the NRI allows NII and the licensees in the IMC to make informed judgements about the strength and weakness of methods that are or may be used in the assessment of UK nuclear plant.

CONCLUSIONS

NII does not prescribe the structural integrity methods to be used by licensees.

Licensees are expected to adopt methods that are suitable for the purpose to which they are applied.

While NII doesn't prescribe methods it does rigorously challenge the licensees approaches to structural integrity.

NII takes account of the opinions of third party outside bodies when considering the use of structural integrity assessment methods on UK nuclear plants.

NII supports research into demonstrating the strengths and weaknesses of new and existing structural integrity methods.

REFERENCES

1. *The Health and Safety at Work etc. Act* (1974 c37).
2. *The Nuclear Installations Act* 1965 (1965 c57).
3. *Title 10, Energy, Code of Federal Regulations*, Part 50, The Office of the Federal Register National Archives and Records Administration, USA.
4. *Safety Assessment Principles for Nuclear Plants*, Health and Safety Executive, 1992.
5. *An Examination of the CEGB's R6 Procedure for the Assessment of the Integrity of Structures Containing Defects*, Advisory Committee on the Safety of Nuclear Installations, Health and Safety Commission, 1989.
6. *Nuclear Research Index 2000/01*, HSE, available at www.hse.gov.uk/nsd/resindex/index.htm, 2000.

CHAPTER 3

The Regulator's View: Non-Nuclear Applications

Harry Bainbridge
Technology Division, Health and Safety Executive

1. INTRODUCTION

In applying the Pressure System Safety Regulations to conventional non-nuclear pressure equipment, the requirements to ensure structural integrity are the same as for nuclear plant. Can you find the defect, can you size it accurately, how do you assess if it could give rise to danger, and are the input values that are used in the assessment correct? This paper discusses the work that Technology Division of the HSE has commissioned in the areas of non-destructive testing (NDT), assessment methodologies, and material properties with regard to conventional pressure equipment, geometry and thickness.

2. NON-DESTRUCTIVE TESTING

The main factors in the NDT of pressure components are:

Selection of appropriate NDT Technique
Defect detection
Coverage
Defect type
Orientation
Component geometry
Location
Sizing tolerance

After discussions with industry, a major project was initiated.

2.1. PROGRAMME FOR THE ASSESSMENT OF NON-DESTRUCTIVE TESTING IN INDUSTRY (PANI)

The nuclear industry has carried out a number of studies into the detection and accuracy of NDT in typical section thickness found in that industry (e.g. the PISC series) but little had been applied to the thicknesses and geometries found in conventional pressure plant. A research project 'Programme for the Assessment of NDT in Industry' (PANI) was set up with the Inspection Validation Centre of AEA Technology (now Serco Assurance). This HSE project was unusual as it was managed by a steering committee, which included the HSE project officer. This committee had an independent chairman with members from the steam-raising, oil-production, chemical-production, vessel manufacturers, insurance companies, NDT providers, gas and distribution industries.

A survey was carried out in these industries as to which NDT technique should be investigated, and manual ultrasonic inspection applied to ferritic components was selected as the research project technique. Two types of in-service defects caused the most concern, wall loss through erosion/corrosion and cracking. These subdivided into erosion/corrosion on pipe bends and root welds and environmental and fatigue cracking at welds. Five specimens were designed and manufactured with a total of 25 defects of different size/type. The specimens were samples of two types of nozzle design, a section of pipework, and shell tee and butt welds. In addition an ex-service Cochrane Wee Chieftain Packager Steam Boiler was obtained that had been taken out of service due to the detection of cracking. These were used for a round robin exercise, where 20 operators from a number of companies performed inspections based either on a supplied procedure, or their own company's procedure. While the project was not specifically aimed at investigating human factors, the Committee did not want the exercise to be a straightforward bench test. A frame was constructed which held the specimens at similar locations to that found in industry. For example the pipework specimen was positioned close to the floor but with ample space to allow the inspection. Generally, the operators attended the IVC laboratory for three days. The inspections were supervised by IVC staff to ensure that no collaboration could occur during the trial.

The results of the inspection of the five manufactured specimens was an overall detection rate of 56% (See Fig. 1), with one defect being detected by just one operator (4%). Defects associated with unfused lands were particularly difficult to detect.

The inspections tended to oversize small defects, and undersize larger defects (See Fig. 2). In the case of the butt weld, the detection rate was 99% for two defects 3 mm high and one defect 8.3 mm high (one third wall thickness). All three defects were measured in the range of 2 to 10 mm, and 75% of the inspectors

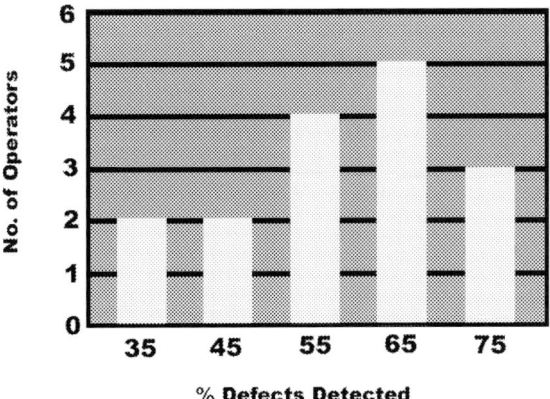

Fig. 1 Operators' performance in detecting defects with ultrasonics in manufactured test pieces.

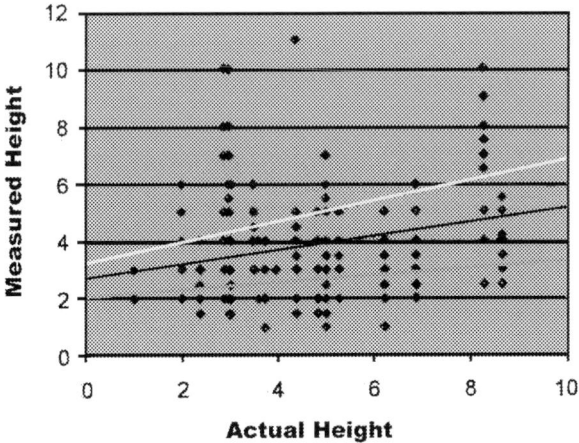

Fig. 2 Plot of measured defect height against actual defect height for the manufactured test pieces.

undersized the larger defect. If a technique accuracy of ± 3 mm is considered, 25% of inspectors undersized outside that range. Analysis showed that inspectors qualified for critical sizing did not achieve better results than the others.

In the boiler operation two thirds of the inspectors reported defects in the furnace tube, but only two operators correctly sentenced the defect that had resulted in the removal from service.

The object of this project was to take a snapshot of the NDT carried out in industry. It illustrated that using qualified NDT inspectors with procedures complying with the national standard did not guarantee an inspection would give the required accuracy that would confirm the integrity of the component.

The Steering Committee continued to meet and produced a document, which listed the factors, which will improve the inspection. This has now been issued as an HSE document 'Best Practice for the Procurement and Conduct of Non-Destructive Testing, Part 1: Manual Ultrasonic Inspection'.

2.2. INSPECTION INDEX

While PANI identified that the detection and sizing of defects was not to the standard expected, in most in-service inspections (ISI), 100% coverage of welds etc. is not carried out. Typical ISI coverage can vary from 5 to 100% depending on the requirements of the written scheme of examination (WSE). In many cases the inspection coverage has been selected from prior experience of that plant or similar plant. A small project looked at attempting to evaluate the amount of coverage on a more scientific basis. Actual inspection data from a vessel was evaluated and it became clear that certain assumptions would have to be made. The defect content in a weld would have to be assumed to be randomly distributed throughout the weld, thus allowing a set of curves shown in Fig. 3 to

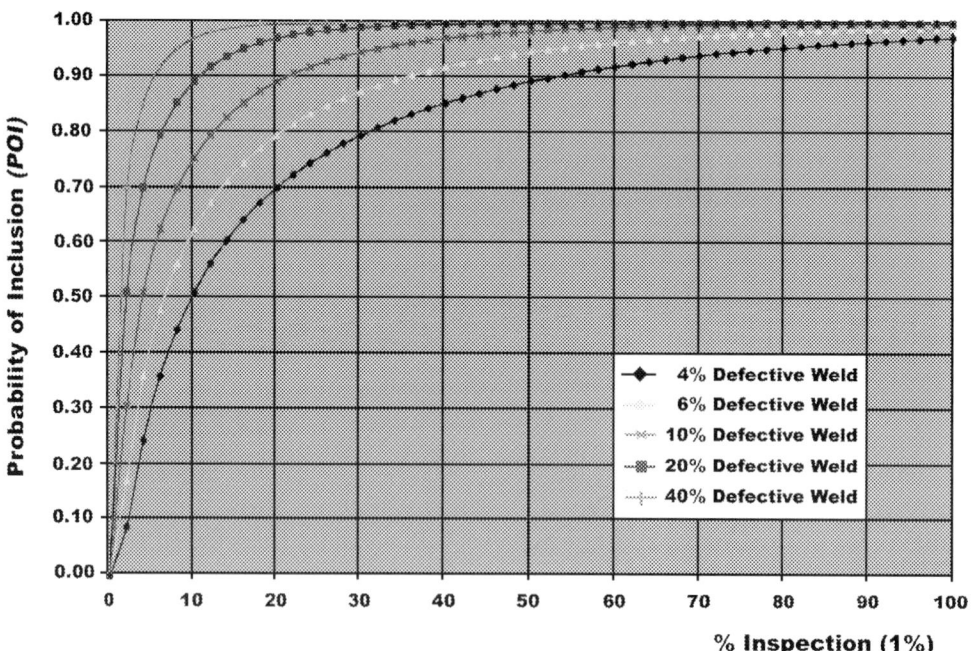

Fig. 3 Probability of including a defective part of the weld given a certain percentage level of inspection.

be generated based on statistical distributions depending on the percentage of defects in the weld, and the length of weld inspected. This gives the probability of inclusion (PoI) of inspection, i.e. the probability that the ultrasonic probe will be placed over a defective section of the weld during the inspection. Figure 4 shows the probability of detection (PoD) curve for manual ultrasonic inspection. Knowing the tolerable defect size will identify the PoD, and this multiplied by the PoI gives an overall inspection index. The inspection index is therefore the overall probability of placing the ultrasonic probe over the defect and detecting it. In the application of this procedure to the inspection of LPG spheres, additional factors are included based on a known population of defects. No guidance has been specified for the required value of the inspection index, as it is acknowledged that the figure will vary depending on the inspection technique and the size of the critical defect. It was considered that a value below 50% would indicate concerns regarding the validity of the inspection to confirm structural integrity.

If this type of sample inspection detected defects, the inspection coverage should be increased until a satisfactory assessment of the integrity of the vessel could be made.

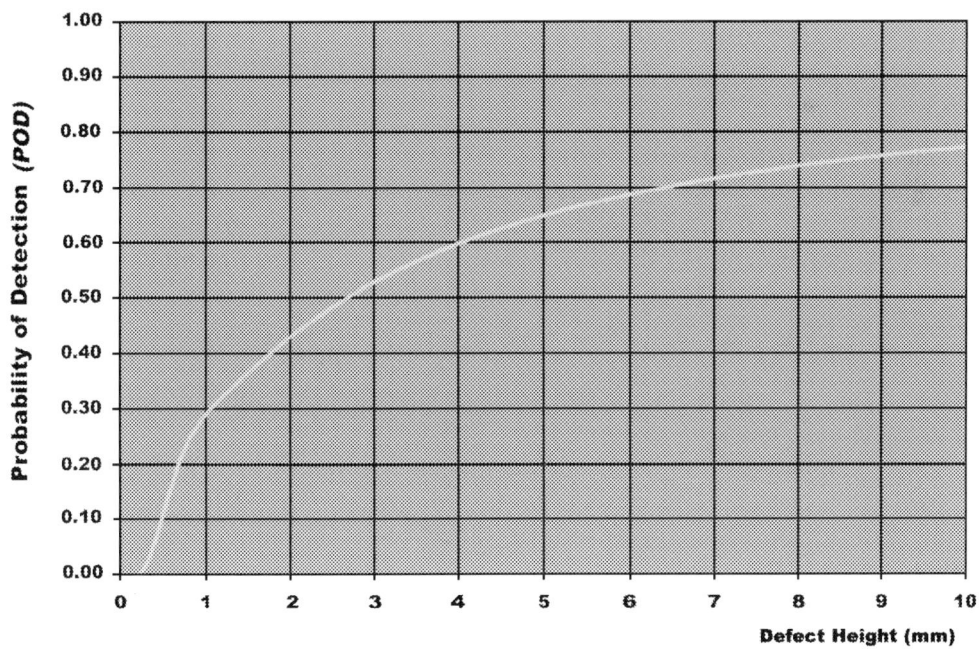

Fig. 4   The lower bound probability of detection vs defect height.

3. ASSESSMENT METHODOLOGIES

For many years the CEGB R6 code has been used for fitness for purpose safety cases in the nuclear industry. For conventional plant, BS PD 6493 has been more widely used for engineering critical assessments (ECA). A new structural integrity document has recently been issued by British Standards Institution BS 7910. As well as incorporating some of the features of the older BS PD 6493 methodologies, BS 7910 contains many aspects of the more state-of-the-art defect assessment methods, and in particular those of R6. Of paramount importance in all ECA methods is the fracture assessment methodology or evaluating critical crack size. A joint HSL/AEAT project[1] was commissioned which assessed four ECA codes, PD 6493, BS 7910, R6 and ASME XI.

The project was divided into three stages. The first action of stage 1 was a literature review to assess whether any comparative studies had been previously reported which could be used to supplement the information generated by this project. Next a series of scoping (benchmark) defect assessment calculations were performed for a number of relevant case histories involving the failure of pressure vessels in service using BS 7910 (level 2) and R6 (Option 1, Category 1). Initial calculations suggested that four of the cases would be suitable for further engineering critical assessments.

For the second stage assessments, the four cases were selected for further analysis using BS 7910 (levels 1 and 2) the 1991 version of PD 6493 (levels 1 and 2) (option 1, category 1) and ASME XI (1992) were:

Case 1 – A reactor pressure vessel (RPV) containing circumferential through-wall and surface cracks in a cylindrical vessel. This case was selected due to the circumferential orientation of the cracks. Assessment of a given crack size was specified (crack length 145 mm for both surface and through thickness cases, crack depth = 8 mm for surface crack, wall thickness = 16 mm).

Case 3 – 'Boiler H' containing an axial through-wall crack in a cylindrical vessel. This case was selected as it has an extremely long and deep crack. Assessment of a given crack size was specified (crack length = 570 mm, wall thickness = 10 mm).

Case 5 – 'Sphere B' containing surface and embedded cracks in a plant spherical vessel. This case was chosen as it contained an embedded defect and the vessel had a different geometry from the other cases. Limiting crack sizes were evaluated.

Case 8 – Risley Moss V2 pressure vessel containing an axial through-wall crack in an experimental cylindrical vessel. This case was chosen as fracture resulted in brittle failure. Limiting crack sizes were evaluated for through-wall and surface axial defects.

Stage three of the project considered the effects of residual stresses for two cases and the effects of weld mis-match for one case. The main findings arising from the assessments were:

(i) In case 5 the differences in the K_r values shown in the failure assessment diagram (FAD) between PD 6493 and BS 7910 methods arise solely due to the fact that the fracture toughness is derived from different $C_V - K_{IC}$ correlations. PD 6493 (based on lower bound experimental data) utilises fracture toughness values derived from either the Charpy energy absorption obtained at the operating temperature or the temperature difference, whichever is the lower. BS 7910 uses Charpy energy-fracture toughness correlations for both upper and lower shelf/transition regions which are a function of the section thickness. In this case for a Charpy impact energy of 40 J at $-20°C$, using BS 7910, a toughness of 80 MPam$^{0.5}$ was obtained, whereas a fracture toughness of 43 MPam$^{0.5}$ was obtained using PD 6493. The K_r values obtained using the latter are therefore much more conservative, leading to smaller tolerable defect sizes. There were no differences in the limit load solutions.

(ii) There are differences in stress intensity factor and limit load (reference stress) solutions between those given in PD 6493 document, those given in the BS 7910 document and those recommended to be used in the R6 assessments. These result in differences in the calculated L_r and K_r values giving rise to differences in safety margins (See Figs 5 & 6).

(iii) The method of dealing with the secondary stress is identical between BS 7910 level 2 and R6. However, the study has shown that significant differences can occur due to interpretation differences by the assessors undertaking the calculations.

(iv) When evaluating the L_r (or S_r) parameter for assessing circumferential defects in a cylindrical pressurised vessel, account should be taken for the possibility of plastic collapse being due to the action of the hoop stress on the un-cracked section. This aspect is not fully explained in any of the procedures considered.

(v) R6 calculations carried out for through thickness defects for the RPV case 1 indicated that most benefit from taking weld strength mis-match into account could be obtained for undermatched weld conditions as opposed to the overmatched conditions.

4. MATERIAL PROPERTIES

With many pressure vessels in the UK approaching the end of their design life, the use of an Engineering Critical Assessment (ECA) as a means of demonstrating

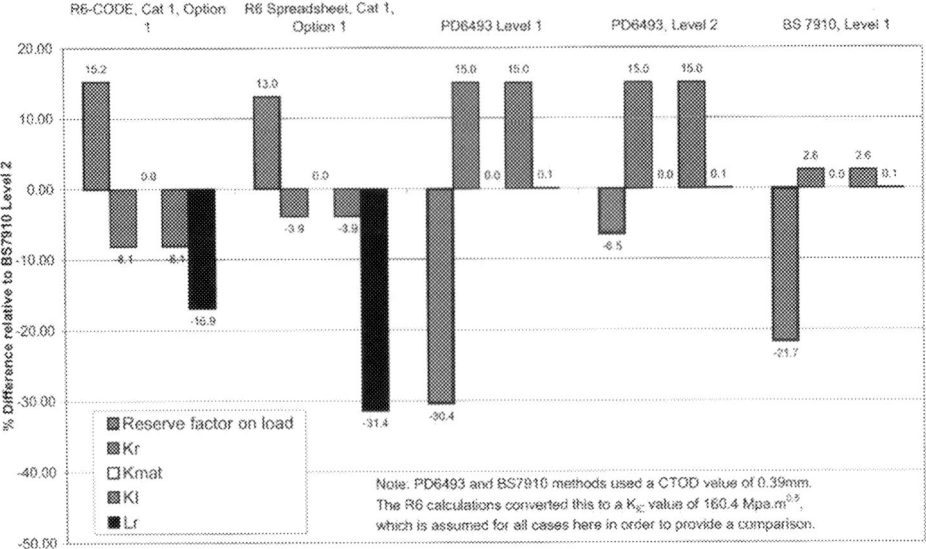

Fig. 5 Case 3: boiler H through-thickness defect, operating pressure, comparison of calculated values relative to BS 7910 level 2.

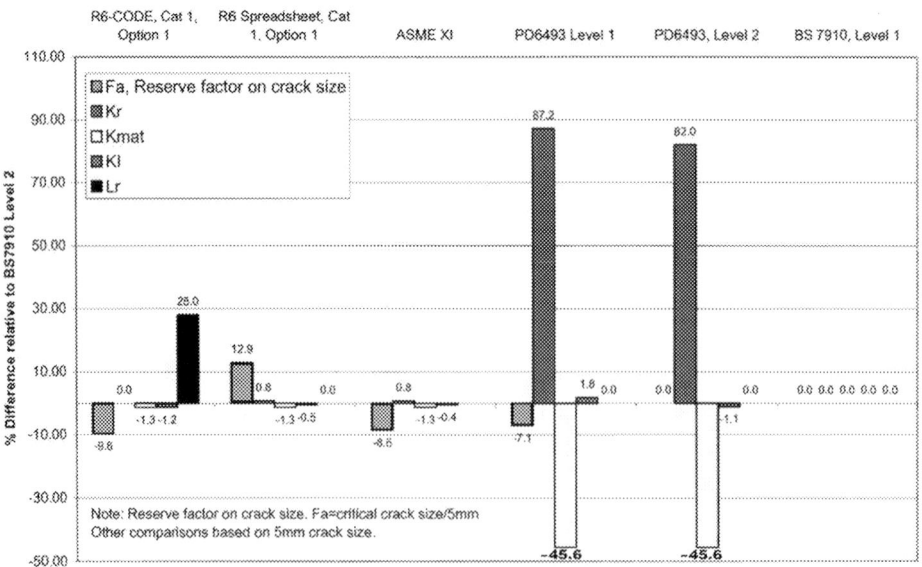

Fig. 6 Case 5: sphere B surface defect, operating pressure, comparison of calculated values relative to BS 7910 level 2.

fitness for purpose and justifying life extensions is used. In many cases the mechanical properties of the material are an unknown factor. In some cases Charpy values have been obtained during manufacture for some or all of the: parent plate, weld metal and heat affected zone (HAZ). These values then have to be converted into fracture toughness at the correct temperature of operation to allow the ECA's to be carried out. This problem identified the requirement for the development of methods which allow the derivation of mechanical properties without substantially damaging the vessel.

Developments in techniques for extracting small 'boat' or 'scoop' samples for the power generation industry promoted a drive to obtain meaningful mechanical properties from these small samples.

4.1. THE DERIVATION OF MATERIAL PROPERTIES FROM SMALL SCALE PUNCH TEST

A small project[2, 3] was commissioned at the Health and Safety Laboratories at Sheffield to evaluate a procedure for obtaining meaningful properties from the scoop samples of BS 1501-161 28A material.

A small disc punch testing procedure was developed which demonstrated that consistent load displacement data could be generated. Results of this work indicated that:

(i) Yield and tensile stress data from uniaxial tensile tests were compared with the punch test data. An empirical relationship between punch test data and tensile stress was explored, and a reasonable correlation was found. The correlation with yield stress was less satisfactory.

(ii) A partial punch test ductile brittle transition (See Fig. 7) was generated with a downward shift of 170°C from that reported from Charpy specimens. A comparison with published data indicated that the transition shifts were sensitive to specimen and rig geometry, and were material specific.

4.2. MINIATURE SPECIMEN FRACTURE TOUGHNESS TESTING

4.2.1. *Pilot Study*

With the considerable temperature shift of 170°C found in the punch test technique, a pilot study into other small specimen techniques was commissioned with HSL and AEA Technology (now Serco Assurance). The object of the project was to establish a co-operative validation of testing procedures within the two organisations, and secondly, to investigate the feasibility of using small scale fracture toughness specimens to define a ductile-to-brittle transition temperature (DBTT) curve. A general engineering grade 50D steel of BS 4360 was used to

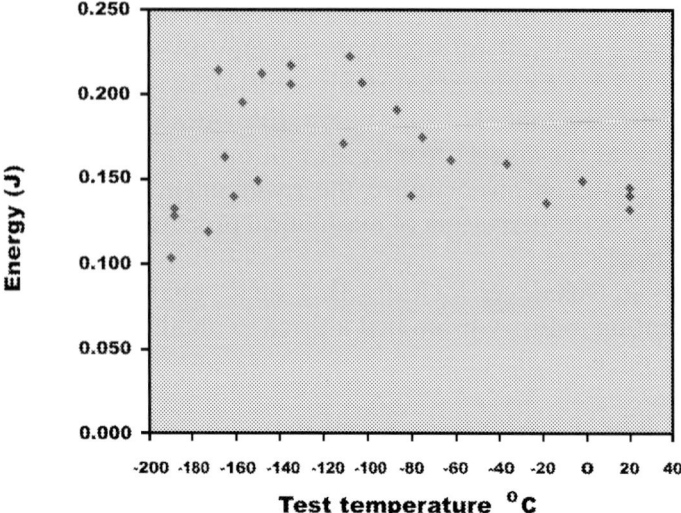

Fig. 7   Punch test transition data.

produce Compact Tension samples in 25 and 5 mm widths. The results (See Figs 8, 9 and 10) provided encouragement for the use of small specimens in obtaining realistic fracture toughness data, with a temperature shift in the DBTT when going from 25 mm specimen thickness to 5 mm thickness being in the order of 50°C.

4.2.2. *Joint Industry Project on Miniature Fracture Toughness Testing*

Following the pilot study, a comprehensive research project was commissioned with the HSL/AEAT partnership to evaluate miniature toughness testing in detail. Three divisions of the HSE and two external companies (Advantica and Rolls-Royce) are involved in this project. Two nuclear quality steels (A533B-1 and A508-CL) which already had a considerable data base regarding toughness testing were used to evaluate specimen size and type, and then parent and weld material from an Advantica HPSV will be evaluated. This work continues.

5. CONCLUSIONS

Technology Division of the HSE has commissioned a series of research projects to study the factors which are used to demonstrate the structural integrity of conventional pressurised plant. This work will continue to reflect industry's trends.

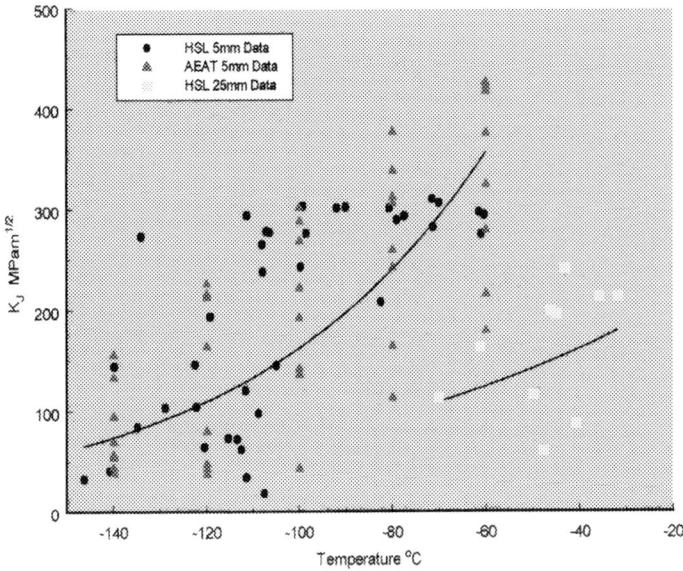

Fig. 8 Cleavage transition data for HSL and AEAT specimens.

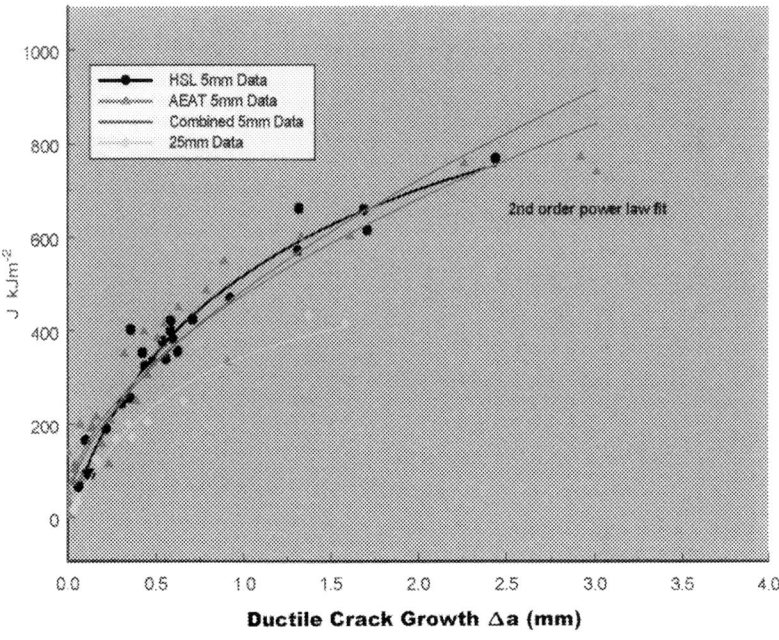

Fig. 9 Resistance curve data for HSL and AEAT specimens.

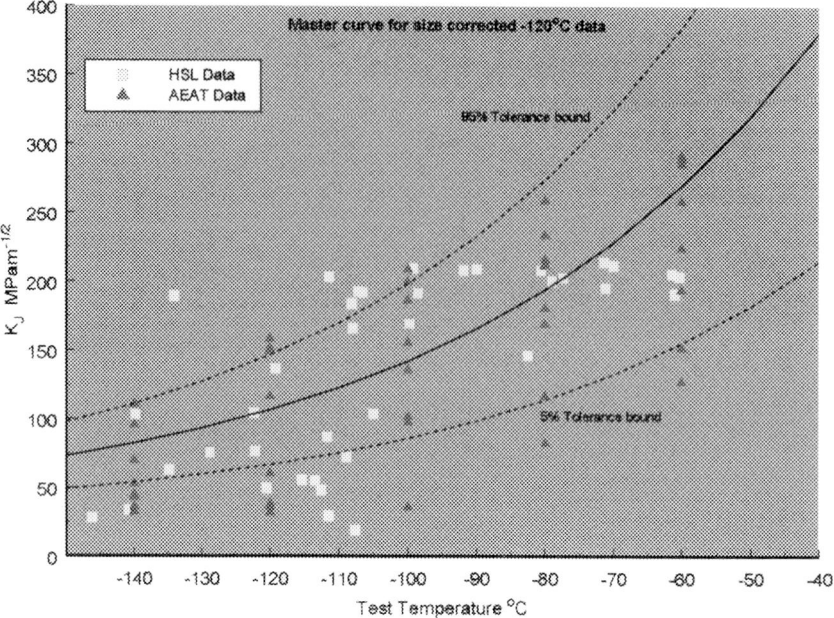

Fig. 10   Toughness transition data for 5 mm CT specimens.

REFERENCES

1. J. K. Sharples, K. May, W. Geary, D. M. Shuter, H. Bainbridge and P. Smith, 'Benchmark Comparisons of Fracture Assessment Methods Utilised in UK Industry', *ASME PVP Conference*, Atlanta, USA, July 2001.
2. D. M. Shuter, J. Dutton and W. Geary, 'The Derivation of Materials Properties Data from Small Scale Punch Tests', *Int. Conf. on Mechanisms and Mechanics of Damage and Fracture*, Vol III, 1996, p. 2205.
3. W. Geary and J. T. Dutton, 'The Prediction of Fracture Toughness Properties from 3 mm Diameter Punch Discs', *Small Specimen Test Techniques*, ASTM STP 1329. W. R. Corwin, S. T. Rosinski and E. van Walle eds, American Society for Testing and Materials, 1997.

CHAPTER 4

SINTAP and R6 Developments: Progress on Codes

A. C. Bannister and S. E. Webster

Corus Research, Development & Technology, Swinden Technology Centre, Moorgate Road, Rotherham, S. Yorks S60 3AR, UK

R. A. Ainsworth

British Energy Generation Ltd, Structural Integrity Branch, Barnett Way, Barnwood, Glos GL4 3RS, UK

ABSTRACT

A co-operative European project on Structural Integrity Assessment Procedures for European Industry (SINTAP) commenced in 1996. The project involved participants from nine European countries. Following a review of existing fitness-for-purpose assessment procedures, and research on a selective number of topics, SINTAP was completed in 1999 with the production of a new flaw assessment procedure. This paper describes some of the developments made during the SINTAP project. Progress in the areas of fracture toughness estimation for ferritic steels in the ductile-to-brittle transition region, compatibility of failure assessment diagram (FAD) and crack driving force (CDF) methods, and allowance for strength mismatch effects are highlighted. Then, an overview of the SINTAP procedure, the various levels of analysis within it and the hierarchical nature of these levels, is presented.

Building on the SINTAP project and other developments in fracture assessment methods, a major new revision (Revision 4) has recently been produced to the R6 defect assessment procedure maintained by the UK nuclear power industry. The last major revision to R6 was in 1986 although there have been a significant number of additions to the document since that time. The structure of R6 Revision 4 is based on that of the SINTAP procedure and consists of five chapters dealing with: basic procedures; inputs to basic procedures; alternative approaches, compendia; and, validation. Furthermore, some new FADs have been introduced in R6 Revision 4. These are also based on SINTAP developments. This paper describes the contents of R6 Revision 4 and in particular the new features introduced in this major revision to the document.

In addition to the SINTAP and R6 developments, there has recently been a number of developments worldwide in fracture assessments codes. These include British Standards developments leading to the British Standards Guide BS7910 which replaces the Published Document PD6493, American Petroleum Institute developments leading to API579, developments leading to the French Nuclear Code RSE-M and developments by the Japanese Society of Mechanical Engineers (JSME). These worldwide code developments are summarised. Finally, the paper discusses progress within Europe towards production of a European standard for fitness-for-purpose assessments.

1. INTRODUCTION

Structural integrity procedures are being increasingly used to analyse the safety of engineering structures. While analytical flaw assessment methods have been developed over a number of years,[1] the increased use of structural integrity evaluations has led to a demand for removal of empiricism and conservatism. Recently, this has led to a number of developments worldwide. In this paper attention is focussed on development in two codes. First, Section 2 describes a co-operative European project on Structural Integrity Assessment Procedures for European Industry (SINTAP) and some of the results from the project.[2] Secondly, a major new revision to the R6 defect assessment procedure,[3] maintained by the UK nuclear power industry, is discussed in Section 3. Finally, the paper briefly summarises other worldwide developments in Section 4.

2. SINTAP DEVELOPMENTS

The SINTAP project[2] was co-ordinated by Swinden Technology Centre, Corus (UK) with the five principal tasks being led by Task Leaders as summarised below.

Task 1: Mismatch – Leader: GKSS
To quantify the behaviour of strength mismatched welded joints and to provide recommendations for their treatment in a procedure.

Task 2: Failure of Cracked components – Leader: British Energy
To extend the understanding of the behaviour of cracked components in the specific area of constraint, yield/tensile ratio, prior overload, leak-before-break, stress intensity factors and limit load solutions.

Task 3: Optimised Treatment of Data – Leader: VTT
To provide an industrially applicable method for a reliability-based defect assessment procedure.

Task 4: Secondary Stress – Leader: IdS
The determination and validation of the most appropriate method of accounting for residual stress, including a compendium of residual stress profiles.

Task 5: Procedure Development – Leader: Corus
Development and validation of the procedure.

Each task comprised a number of subtasks, the first being a comprehensive collation of existing data, procedures and codes. This was then followed by experimental and numerical work to cover omissions and to validate the approaches and assumptions relevant to the particular study area.

It is not possible in this paper to cover all these aspects in detail. Instead the subsections below highlight some particular developments with further developments being described in Section 3 in the context of their inclusion in R6.

2.1. TREATMENT OF TOUGHNESS DATA

Within the SINTAP Procedure, a methodology is included in which one material-specific toughness parameter, K_{mat}, together with its probability density distribution is defined.[4] For assessment against brittle fracture, the approach uses a 'Master Curve' method to describe the temperature dependence of fracture toughness. Indirect methods based on Charpy impact data can also be used. For ductile tearing, standard methods for deriving characteristic values of fracture toughness as a function of ductile crack growth are given and these are not described here.

For indirect determination from Charpy impact energy, no single correlation can be applied to all parts of the toughness transition curve. In the SINTAP procedure the following options are available:

(i) A lower bound correlation for brittle (lower shelf) behaviour;
(ii) A statistical method for the transition regime (the 'Master Curve');
(iii) A lower bound correlation for the ductile (upper shelf) behaviour.

For example, the 'Master Curve' concept is based on a correlation between the Charpy 28J temperature and the temperature $K_{mat} = 100\,\text{MPa}\,\text{m}^{0.5}$. The relationship is then modified to account for the required failure probability, thickness effect and the shape of the fracture toughness transition curve. The transition curve for fracture toughness in the transition regime can be defined as

$$K_{mat} = 20\{11 + 77\exp(0.019[T - T_{28J} - 3°C])\}(25/B)^{1/4}\{\ln(1/[1 - P_f])\}^{1/4} \quad (1)$$

where T = temperature at which the toughness K_{mat} applies (°C); $T_{28J} = 28/27$J Charpy transition temperature (°C); B = specimen thickness or flaw length (mm) and P_f = probability of fracture.

The SINTAP procedure for estimation of brittle fracture toughness involves a series of stages. The first stage is to check that all the data meet the acceptance criteria of the relevant testing standard. The fracture toughness values are then converted into equivalent stress intensity factors with $\text{MPa}\,\text{m}^{0.5}$ units. Next, the specimen capacity limit (K_{climit}) is determined from:

$$K_{climit} = (Eb_o\sigma_y/30)^{0.5} \tag{2}$$

where σ_y is the yield or proof stress and b_o is the initial ligament below the notch in the specimen.

Equation (2) ensures that fracture occurs under small-scale yielding and results from specimens that exceed this limit are censored: a censoring parameter, δ, is set at 0 and the fracture toughness set at K_{climit}. Specimens that do not fracture are also censored ($\delta = 0$). Results from other specimens are not censored and δ is set at 1 for each specimen.

Next, the fracture toughness values are corrected to a reference thickness of 25 mm, K_{c25}. The values of K_{c25} and their associated censoring parameters are then used to make the first estimate of toughness, K_o:

$$K_o = 20 + \left[\frac{\sum\limits_{i=1}^{N}(K_{c25i} - 20)^4}{\sum\limits_{i=1}^{N}\delta_i}\right] \tag{3}$$

where N is the number of results. Note that the procedure uses all the test results and that K_o is biased towards the uncensored data ($\delta = 1$).

Step 2 in the procedure, (referred to as 'lower tail estimation') involves censoring all data above the 50th percentile of the distribution, i.e. setting $\delta = 0$ for all the values above the 50th percentile. This ensures that the estimate of K_o is biased towards the lower tail of the toughness distribution so as to include results from specimens containing local brittle zones (LBZs). Results from specimens which do not contain LBZs and that are likely to give high fracture toughness values tend to be excluded by Step 2. The censored values are assigned the median value of toughness.

$$\bar{K}_{c25} = 20 + (K_o - 20)0.91 \tag{4}$$

After censoring, K_o is re-estimated using eqn (3). However, since both eqns (3) and (4) contain K_o, the procedure is iterative with K_o and $\bar{K}_{c25}$ being continually adjusted until a consistent K_o is obtained.

The final step, Step 3 (referred to as 'minimum value estimation'), requires an estimate of K_o using the minimum fracture toughness value in the data set. The values of K_o from each of the three steps are now compared and the lowest determined from the three steps taken, except when K_o from Step 3 is not less than 90% of the lower of K_o from Steps 1 or 2. Where the value from Step 3 is less than 90% it would be conservative to use K_o from Step 3. However, since the procedure is highlighting an outlier, a judgement has to be made as to its significance. If the dataset is large and the fit to the assumed distribution is good, then the result from Step 3 can be treated as representing an anomaly and can be ignored. It is then reasonable to assume that the lower of Steps 1 or 2 are describing the fracture toughness distribution. However, if the data set is small, it would be unreasonable and unconservative to ignore Step 3. If the result from Step 3 is considered to be unsatisfactory then further testing should be conducted to better define the lower tail of the fracture toughness distribution.

The procedure described above can also, in principle, be applied to data in the fracture toughness transition regime.

2.2. COMPATIBILITY OF FAD AND CDF APPROACHES

The two principal fracture assessment methods that are used for analysis are the Failure Assessment Diagram (FAD) and the Crack Driving Force (CDF).[1] The former method is used in approaches such as R6,[3] BS 7910[5] and API 579;[6] whilst the CDF approach is favoured in ETM[7] and GE-EPRI[8] procedures.

The basis of both approaches is that failure is avoided so long as the structure is not loaded beyond its maximum load bearing capacity defined using fracture mechanics criteria and plastic limit load analysis. The CDF approach involves comparison of the loading on the crack tip (often called the crack tip driving force) with the ability of the material to resist fracture (defined by the material's fracture toughness or fracture resistance). Crack tip loading must, in most cases, be evaluated using elastic–plastic concepts and is dependent on the structure, the crack size and shape, the material's tensile properties and the loading. In the FAD approach both the comparison of the crack tip driving force with the material's fracture toughness and the plastic load limit analysis are performed at the same time. Whilst both approaches are based on elastic–plastic concepts, their application is simplified by the use of only elastic parameters.

A critical review of these procedures was carried out within SINTAP and this demonstrated that there is relatively little difference between results obtained. Although FAD and CDF routes represent two different calculation methodologies, the underlying principles remain the same.[1] It is only in their development, through different simplifications, that they diverge.

In the following description, the term 'yield stress' is used as either the yield stress in the case of discontinuous yielding, or the 0.2% proof stress for continuous yielding. Differentiation between these two parameters is only made where necessary. In the CDF method, calculations are made of an applied parameter such as J-integral or crack opening displacement that characterises the state of stress and strain ahead of the tip of a crack in a component. J is estimated as

$$J = J_e[f(L_r)]^{-2} \qquad (5)$$

where J_e is the elastic value of the J integral which can be deduced from the stress intensity factor K_1. A number of options for describing the function $f(L_r)$, are described in Section 3. In particular, the functions $f(L_r)$ within SINTAP approach for power-law materials were based directly on the equations in ETM[7], while others were based directly on those in R6[9]. The load ratio

$$L_r = F/F_Y \qquad (6)$$

where F is the applied load and F_Y is the limit load defined from the yield strength.

It can be seen that J can be estimated from eqns (5, 6) provided the applied load F is known, a stress intensity factor solution is available (note K_1 is proportional to F and depends on the geometry and flaw size) and a limit load solution is available (note F_Y is proportional to yield stress and depends on geometry and flaw size). In the CDF method, fracture is conceded when J exceeds a material property value, J_{mat}, which is related to fracture toughness, K_{mat}.

In the FAD method, two parameters are calculated. One is the load ratio L_r already defined by eqn (6). The second is a ratio K_r, defined by

$$K_r = K_1/K_{mat} \qquad (7)$$

for simple primary loading.

Once these two parameters have been calculated, fracture is avoided if the point (L_r, K_r) is within a region defined on a failure assessment diagram. The failure avoidance region is given by the failure assessment curve

$$K_r = f(L_r) \qquad (8)$$

and a cut-off

$$L_r = L_r^{max} \qquad (9)$$

Manipulation of eqns (5–8) shows that the condition $K_r \leq f(L_r)$ is equivalent to $J \leq J_{mat}$ so that the CDF and FAD representations within the SINTAP procedure are fully compatible. The criterion (9) is imposed to provide a plastic collapse limit with L_r^{max} dependent on the strain hardening characteristics of the material. This limit is imposed in both the CDF and FAD methods.

2.3. TREATMENT OF STRENGTH MISMATCH

The SINTAP method for weld strength mismatched structures[10] was based on existing proposals for modified R6 and ETM methods. The SINTAP method for strength mismatch has four levels, depending on the quality of tensile information.

A central feature of the procedure is that F_Y in eqn (6) is replaced by a mismatch limit load, F_{YM}, which depends on the yield strengths of both base and weld materials, σ_{yB} and σ_{yW} respectively. As part of the SINTAP project, a collation of solutions for a number of geometries, flaw sizes and mismatch ratios

$$M = \sigma_{yW}/\sigma_{yB} \tag{10}$$

was produced. Then a number of options for defining the function $f(L_r)$ in eqn (5) were developed.

The detailed equations for $f(L_r)$ will be illustrated here using the failure assessment diagram (FAD) approach only for Level 2 (see Section 2.4) and for two cases within that level: (1) when both base and weld material do not exhibit Lüders strain; (2) when both materials do.

In the former case the following equation can be used for $0 \le L_r < L_r^{max}$,

$$f(L_r) = (1 + 0.5L_r^2)^{-1/2}[0.3 + 0.7\exp(-\mu_M L_r^6)] \tag{11}$$

where $\mu_M = \min\left[\dfrac{(M-1)}{(F_{YM}/F_{YB}-1)/\mu_W + (M - F_{YM}/F_{YB})/\mu_B}, 0.6\right] \tag{12}$

with $\mu_W = \min[0.001E_W/\sigma_{yW}, 0.6], \mu_B = \min[0.001E_B/\sigma_{yB}, 0.6] \tag{13}$

Here E_W, E_B are Young's moduli for weld and base materials, respectively and F_{YB} is the limit load assuming the component is made wholly of base material.

The cut-off L_r^{max}, is determined from

$$L_r^{max} = \frac{1}{2}\left(\frac{1+0.3}{0.3 - N_M}\right) \tag{14}$$

where the strain hardening exponent for the mismatched component, N_M, is estimated from

$$\frac{N_M = (M-1)}{(F_{YM}/F_{YB}-1)/N_W + (M - F_{YM}F_{YB})/N_B} \tag{15}$$

The hardening exponents for the weld and base material, N_W and N_B, are estimated from

$$N_W = 0.3(1 - \sigma_{yW}/\sigma_{uW})$$
$$N_B = 0.3(1 - \sigma_{yB}/\sigma_{uB}) \tag{16}$$

where σ_u denotes the ultimate tensile strength.

For $L_r > 1$, the following equation is used up to $L_r = L_r^{max}$;

$$f(L_r) = f(1)(L_r)^{(N_M-1)/2N_M} \tag{17}$$

where $f(1)$ is determined from eqn (11).

When both materials exhibit Lüders strain, the function $f(L_r)$ is

$$f(L_r) = (1 + 0.5L_r^2)^{-1/2} \qquad 0 \le L_r < 1 \tag{18}$$

At $L_r = 1$, the function $f(L_r)$ is taken as discontinuous and reduces to the value $f(1)$, which is dependent on the extent of the Lüders strain:

$$f(1) = \left(\lambda_M + \frac{1}{2\lambda_M} \right)^{-1/2}$$

$$\lambda_M = \frac{(F_{YM}/F_{YB} - 1)\lambda_W + (M - F_{YM}/F_{YB})\lambda_B}{(M - 1)}$$

$$\lambda_W = 1 + \frac{E_W \Delta \varepsilon_W}{\sigma_{yW}} ; \Delta \varepsilon_W = 0.0375 \left(1 - \frac{\sigma_{yW}}{1000} \right)$$

$$\lambda_B = 1 + \frac{E_B \Delta \varepsilon_B}{\sigma_{yB}} ; \Delta \varepsilon_B = 0.0375 \left(\frac{1 - \sigma_{yB}}{1000} \right) \tag{19}$$

where $\Delta \varepsilon$ represents the estimated extent of Lüders strain; the subscripts, B and W, denote the properties of the base and weld materials. For $L_r > 1$, eqn (17) is again used up to $L_r = L_r^{max}$ with N_M defined by eqn (15).

Validation of the mismatch procedure has been addressed by finite element modelling and an experimental programme. In the latter two different materials, having different yield and ultimate strength values, were produced by heat-treating an A533B-1 steel. The yield strength of the higher strength material (designated as $M1$) is about 50% higher than that of the lower strength material (designated as $M3$). It should be noted that the $M3$ material exhibits Lüders strain of a length about 0.8%. In the $M1$ material, however, a Lüders plateau was not visible, due to its high yield strength and low hardening capacity at low strains.

Strength mismatched specimens with idealised weldments were produced by electron-beam (EB) welding, resulting in two different strength mismatched specimens: highly over-matched ($M \approx 1.48$) and highly under-matched ($M \approx 0.68$) specimens. Single edge notched specimens in three-point bend with a total of 20% side grooving were produced, having the crack in the centre of the weldment. Two different crack lengths were chosen, $a/W = 0.45$ and 0.65. All specimens failed by extensive ductile tearing, and the maximum loads were obtained from experimental records.

For both overmatched and undermatched specimens, the predicted maximum loads, F_{max}^{pred}, are shown in Fig. 1, together with the measured maximum loads, F_{max}^{exp}, in the test. The prediction using the SINTAP Level 3 (see Section 2.4) FAD is conservative but only by 10%. The Level 2 FAD gives slightly more

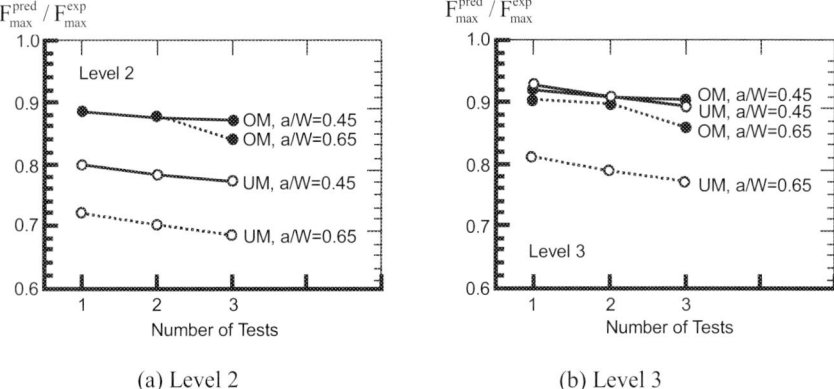

Fig. 1 Comparison of predicted maximum loads with experimentally measured loads for the SINTAP mismatch procedure 2.4 SINTAP Levels.

conservative results. Such results are likely to arise from the conservatism embedded in the Level 3 curve for materials with Lüders strain.

A similar trend has been observed for ductile tearing analyses. The results from this experimental validation strongly support the methodology of the SINTAP method for weld strength mismatch. Further validation details for mismatch and for the homogenous SINTAP procedure are described elsewhere.[10,11]

2.4. SINTAP LEVELS

The SINTAP procedure provides a range of analysis levels enabling advantage to be taken of increasing data quality where data are available, and reflecting the variation in user knowledge and experience. In brief, the levels are as follows:

L0: Default — to be used when only yield stress data are available. The fracture toughness may also be estimated from Charpy impact data as described in Section 2.1 so that this level may be followed when available materials data are limited.

L1: Basic — this level is used when only yield and ultimate tensile stress data are available and in the presence of welds where strength mismatch levels are less than 10%.

L2: Mis-match — this is similar to Level 1 in terms of availability of tensile data but can be applied for strength mismatch levels in excess of 10% and therefore requires data for both parent plate and weld metal to follow the procedures of Section 2.3.

L3: Stress-strain – this is a material specific level which requires full stress–strain curves to be known. It is equivalent to use of the R6 Option 2 failure assessment curve described in Section 3.1 below. Strength mis-match can be addressed but only if stress-strain data are available for both parent and weld metals.

L4: Constraint – this uses fracture toughness estimates, relevant to the crack tip constraint conditions, which may be higher than those for the lower levels. This enables reduced conservatism but required additional test data.

L5: *J*-integral – this level uses full stress-strain data as input to a numerical analysis to determine *J*. This leads to reduced conservatism compared to the lower analysis levels which may be considered as approximate *J*-estimation techniques.

L6: LBB – in this level, the stability and growth of a part-penetrating flaw and the resultant through-thickness flaw in a pressurised component are assessed by a leak-before-break analysis.

3. R6 DEVELOPMENTS

The last major revision (Revision 3) to the R6 defect assessment procedure was produced in 1986.[9] Since that time, there have been numerous additions to the document and revisions to parts of it, but the structure of the procedure has remained unchanged. Following completion of the SINTAP project it was decided to produce Revision 4 of R6.[3] The structure of R6 Revision 4 is similar to that of the SINTAP Procedure and consists of five chapters:

Chapter I   –   Basic Procedures
Chapter II   –   Inputs to Basic Procedures
Chapter III –   Alternative Approaches
Chapter IV   –   Compendia
Chapter V   –   Validation and Worked Examples

The underlying principles of the Basic Procedures of Chapter I and the related Inputs in Chapter II are largely unchanged from previous versions of R6. However, there have been a number of significant changes[12] as follows:

(i) In R6 Revision 3[9] there were three categories of analysis. Category 1 was an initiation assessment. Category 2 was a simplified treatment of ductile tearing analysis and Category 3 was a full treatment. Recognising that ductile tearing analysis has become established since Revision 3 was produced and that the simplified Category 2 approach was rarely used, Category 2 has not been included in Revision 4 and the terminology 'category of analysis' has been removed. Therefore, analyses are simply referred to as 'initiation' or 'ductile tearing'.

(ii) In Revision 3, the stress σ_y used to define the normalised load parameter L_r was variously referred to as the 'yield stress' and the '0.2% proof stress'. In Revision 4, any potential ambiguity has been removed and σ_y is consistently defined as the 0.2% proof stress irrespective of the detailed shape of the stress-strain curve.

(iii) R6 has always distinguished between stresses which contribute to collapse (primary stresses) and those which do not (secondary stresses). Interaction between these types of loading has been included in the definition of K_r which is modified from eqn (7) to

$$K_r = (K_1^p + K_1^s)/K_{mat} + \rho \qquad (20a)$$

where K_1^p and K_1^s are the stress intensity factors for the primary and secondary stresses, respectively, and ρ is a factor covering interactions. In R6 Revision 4, an alternative but equivalent approach has also been included in which K_r is defined as

$$K_r = (K_1^p + VK_1^s)/K_{mat} \qquad (20b)$$

where the factor V now covers interactions. The V approach has been described fully elsewhere[13] and is not discussed further in this paper.

(iv) Some new failure assessment curves have been introduced. These are based on developments within SINTAP[14] and are described in Section 3.1 below.

(v) Some new advice on flaw characterisation had been included as summarised in Section 3.2 below.

The remaining Chapters III–V have had more substantial revisions than the basic procedures and these are discussed in Section 3.3–3.5, respectively.

3.1. THE R6 FAILURE ASSESSMENT CURVES

In Revision 4 of R6, there are a number of failure assessment curves:

- Option 1
- Option 2
- Approximate Option 2
- Option 3
- C–Mn Steels

The last two of these are unchanged from R6 Revision 3, although Revision 4 contains more detailed advice on producing an Option 3 curve, and are not discussed further here. The R6 Option 2 curve is based on detailed stress–strain data being available and is

$$f_2(L_r) = \left[\frac{E\varepsilon_{ref}}{L_r\sigma_y} + \frac{L_r^3\sigma_y}{2E\varepsilon_{ref}} \right]^{-1/2} \tag{21}$$

where ε_{ref} is the strain from uniaxial tensile data at a stress level $L_r\sigma_y$ and E is Young's modulus. Clearly, this curve depends on the strain-hardening behaviour of a material including the length, if any, of a yield plateau (Lüders strain). Following a review of existing procedures within SINTAP, this curve was adopted within the SINTAP procedures and, therefore, has been retained unchanged in R6 Revision 4.

The Option 1 curve of R6 Revision 3[9] was examined within the SINTAP project. This curve is given by:

$$f(L_r) = (1 - 0.14L_r^2)[0.3 + 0.7\exp(-0.65L_r^6)] \tag{22}$$

It was chosen within R6 as an empirical fit, biased towards the lower bound of curves derived from eqn (21) for a range of materials not exhibiting discontinuous yielding.[15] It has widespread validation by comparison with both experimental fracture data and results of numerical analyses. However, it does not reproduce eqn (21) at low L_r. To ensure a hierarchical approach within the SINTAP procedures, an alternative function was, therefore, chosen as an approximation to eqn (22) (the fit is within about 3% for $0 \leq L_r \leq 1$) while reproducing the required behaviour of eqn (21) at low L_r. This equation has been included in R6 Revision 4 as the Option 1 curve

$$f_1(L_r) = (1 + \tfrac{1}{2}L_r^2)^{-1/2}[0.3 + 0.7\exp(-0.6L_r^6)] \tag{23}$$

Based on the work of SINTAP, two new failure assessment curves have been introduced in R6 Revision 4. These are similar to the Level 1 curves of SINTAP and in R6 Revision 4 are termed approximate Option 2 curves for discontinuous hardening materials and for continuous hardening materials.

Materials which show discontinuous yielding may be considered to exhibit a stress-strain curve consisting of three regions:

- elastic behaviour up to the (lower) yield stress, σ_y;
- an increase in strain, $\Delta\varepsilon$, at the stress σ_y without any increase in stress;
- further plastic straining with increasing stress up to the ultimate stress, σ_u.

For this behaviour, the approximate Option 2 failure assessment curve in R6 Revision 4, $f_2^{dy}(L_r)$, is obtained from the Option 2 function of eqn (21) as

$$f_2^{dy}(L_r) = (1 + 0.5L_r^2)^{1/2} \quad L_r 1 \tag{24}$$

$$f_2^{dy}(1) = \left[\frac{\lambda + 1}{2\lambda}\right]^{1/2} \quad L_r = 1 \tag{25}$$

where
$$\lambda = 1 + E\Delta\varepsilon/\sigma_y \tag{26}$$

In the absence of detailed stress-strain data enabling eqn (25) and the Option 2 curve to be evaluated for $L_r > 1$, then $\Delta\varepsilon$ may be estimated from

$$\Delta\varepsilon = 0.0375[1 - \sigma_y/1000] \tag{27}$$

for σ_y in MPa. Application of eqn (27) is restricted to materials for which $\sigma_y < 946$ MPa, for which the estimate of $\Delta\varepsilon$ exceeds 0.2%. The curve for $L_r > 1$ may be estimated from

$$f_2^{dy}(L_r) = f_2^{dy}(1)L_r^{(N-1)/2N} \quad 1 < L_r < L_r^{max} \tag{28}$$

with N estimated from

$$N = 0.3[1 - \sigma_y/\sigma_u] \tag{29}$$

For $L_r > 1$, this corresponds to an estimated power-law fit to the plastic portion of the stress-strain curve with the value of the power-law exponent, N, obtained from correlations with tensile data for a range of steels.

Note that f_2^{dy} is discontinuous at $L_r = 1$ since $\lambda > 1$. It can be seen that the curves are consistent with the SINTAP mismatch curves presented in Section 2.3.

In the absence of detailed stress-strain data, and where discontinuous yielding is not expected, an approximate Option 2 curve, $f_2^{cy}(L_r)$, is constructed by modifying the Option 1 curve of eqn (23) as follows

$$f_2^{cy}(L_r) = [1 + 0.5L_r^2]^{-1/2}[0.3 + 0.7\exp(-\mu L_r^6)] \quad L_r \leq 1 \tag{30}$$

$$f_2^{cy}(L_r) = f_2^{cy}(1)L_r^{(N-1)/2N} \quad 1 < L_r < L_r^{max} \tag{31}$$

where $f_2^{cy}(1)$ is obtained from eqn (30) for $L_r = 1$, N is obtained from eqn (29) and

$$\mu = \text{Min}[0.001E/\sigma_y; 0.6] \tag{32}$$

For continuously hardening materials, eqs (30–32) have the following features:

(i) They are identical to the Option 1 curve of eqn (23) for $L_r \leq 1$ for materials with $\sigma_y/E \leq 1/600$ and hence $\mu = 0.6$;
(ii) They reduce conservatisms for $L_r \leq 1$ relative to Option 1 for materials with $\sigma_y/E > 1/600$.

(iii) The reduced conservatisms at $L_r = 1$ are consistent with Option 2 which leads to the value $f(1)$ being dependent on σ_y/E, see eqn (21). However, the reduction in conservatism from Option 1 at $L_r = 1$ will be less than that obtained if full stress–strain data are available and Option 2 is used.

Again, they are consistent with the SINTAP curves for mismatch presented in Section 2.3.

3.2. FLAW CHARACTERISATION

Flaw characterisation rules enable naturally complex-shaped and irregularly oriented and distributed defects to be analysed and assessed. Guidance on flaw characterisation in Revision 4 of R6 extends guidance from R6 Revision 3 as described below.

First, the section on flaw orientation has been significantly expanded by incorporating guidance given in API 579.[6] In particular, comprehensive guidance is given on how to project flaws on to reference planes and how to deal with branch cracks.

R6 allows a defect to be recharacterised if an initial assessment indicates that failure of a ligament may not be avoided. On recharacterising a part through defect, when the ligament fails by a ductile mechanism the length at penetration should be increased by an amount equal to the local component thickness to allow for lateral crack growth. The evidence about the necessity for this is conflicting but the length increase recommended is judged to be conservative. This has been confirmed by a recent TAGSI review[16] of recharacterisation. When ligament failure occurs by a brittle mechanism, it is necessary to allow for dynamic effects at the moving crack tip. Consideration of the kinetic energy involved in the process leads to a judgement that a factor of 2 increase on defect length is sufficient to allow for dynamic conditions. A more rigorous analysis of dynamic conditions would invoke crack arrest argument. Dynamic effects are generally small in the fully ductile regime and on the lower shelf, and are maximised in the transition regime. However, care needs to be exercised in the use of recharacterisation rules in the transition regime.[16]

For multiple defects or defects close to a free surface, the interaction rules in R6 Revision 4 incorporate the guidance in BS7910.[5] This guidance is less conservative than that of R6 Revision 3.[9] The revised guidance on interaction effects is also consistent with that given in the SINTAP procedure .

In Revision 3 of R6, it was stated that if multiple flaws did not interact according to the interaction criteria then they could be considered as separate defects with stress intensity factors enhanced as appropriate. A factor of 1.2 on the stress intensity factor was stated as always being sufficient for such enhancement. No such factor is included in BS7910. The TAGSI Fracture Subgroup[17] examined a range of numerical and experimental evidence related to defect interaction rules. For two surface cracks in the same plane, although consideration of the numerical stress intensity factor parameter would indicate enhancement, this was not apparent from experimental evidence. It was considered that the lack of enhancement observed from experimental evidence was due to loss of constraint. Therefore, it is now considered that the inclusion of the enhancement factor in the advice given in Revision 3 of R6 was overly conservative. Thus, such a factor has not been included in Revision 4. However, inclusion of such a factor should be considered if the flaw interaction rules are used in conjunction with constraint-based approaches.

A new section has been introduced in R6 Revision 4 dealing with flaw detection, sizing and non-destructive evaluation aspects. This section is taken from the SINTAP procedure and has been incorporated, since it is recognised that inspection technology plays an important role in the initial characterisation of both single and multiple flaws.

3.3. R6 REVISION 4 CHAPTER III, ALTERNATIVE APPROACHES

The sections in Chapter III may broadly be described as being of three types: refinements to the basic procedures; procedures to reduce conservatisms; and different approaches. The 14 sections are:

III.1   Selection of Alternative Approaches
III.2   Finite-Element Analysis
III.3   *J*-Estimation Approaches
III.4   Sustained Loading
III.5   Mode I, II and III Loads
III.6   Structures Made of C–Mn (Mild) Steel
III.7   Constraint Effects
III.8   Strength Mis-Match
III.9   Local Approach Methods
III.10 Prior Loading and Warm Pre-Stressing
III.11 Leak-Before-Break Assessment
III.12 Crack Arrest
III.13 Probabilistic Fracture Mechanics
III.14 Displacement-Controlled Loading

A number of these sections are largely unchanged from the previous advice in R6 Revision 3.[9] However, there have been some significant changes and these are highlighted below.

Section III.3 on *J*-Estimation Approaches is a new section. It provides comprehensive advice on derivation of an Option 3 failure assessment curve:

$$f_3(L_r) = (J_e/J)^{1/2} \tag{33}$$

where J_e and J are the elastic and elastic-plastic values of J for the same load. It also provides a procedure for presenting results as a crack driving force. The latter advice is based on the SINTAP procedure and is fully consistent with the FAD approach as discussed in Section 2.2 above. Load-order effects are also discussed in Section III.3.

Recent experimental work and related validation analyses have demonstrated that the Revision 3 procedures are adequate for mixed mode loading Therefore, the Revision 3 procedures have been retained in Section III.5. However, the status notes and supporting references have been extensively revised to reflect the new work.

In Section III.7, updated advice to that in Revision 3 on treatment of constraint effects is given. Some validation information in Revision 3 has been removed to Chapter V and revised status notes are given which include references to numerical solutions produced during the SINTAP project.

Extended advice to that in Revision 3 Appendix 16 on strength mismatch effects is described in Section III.8. New validation data in support of the methodology is contained in Chapter V. Additional information is supplied for components containing asymmetric cracks. For consistency with the new approximate Option 2 curves described above, corresponding new curves for mis-match have been added, based on those presented in Section 2.3.

Recent work on leak-before-break has developed advice on reference stress methods for calculating crack opening areas in the presence of plasticity. Section III.11 contains that advice along with the comprehensive advice previously in Revision 3 Appendix 9 on leak-before-break that includes advice for high temperatures where creep effects are important.

Advice on probabilistic fracture mechanics was previously given in Revision 3 Appendix 10. However, there has been significant recent work in this area worldwide with advice contained in the BS7910[5] and SINTAP[2] procedures, for example. This advice does not affect the underlying principles of Appendix 10 but has enabled the new Section III.13 to be extensively revised from Appendix 10 to include new information.

3.4. R6 REVISION 4 CHAPTER IV, COMPENDIA

There are four new compendia in R6 Revision 4 which make the documentation much more self-contained:

IV.1   Limit Load Solutions for Homogeneous Components
IV.2   Limit Load Solutions for Strength Mis-Match
IV.3   Stress Intensity Factor Solutions
IV.4   Welding Residual Stress Distributions

The limit load solutions in Section IV.1 are those in a recently completed compendium produced as part of the R6 development programme. This updated the Miller compendium.[18] A mismatch compendium was produced at GKSS in Germany and finalised during the SINTAP project. This is the basis for Section IV.2.

Stress intensity factor solutions are contained in Section IV.3. This section is not intended to be a replacement for established compendia such as those of Tada[19] and Murakami.[20] Instead it concentrates on solutions for practical component geometries such as cylinders, although solutions are also given for plates and test specimens. Where possible, weight or influence function solutions are given so that stress intensity factors can be derived for complex thermal or residual stress fields.

A SINTAP compendium on welding residual stresses was based on a previous R6 compendium referred to in Revision 3. This has been updated to become Section IV.4 of R6 Revision 4. A similar compendium is contained in BS7910.[5]

3.5. R6 REVISION 4 CHAPTER V, VALIDATION AND WORKED EXAMPLES

This part of Revision 4 is also new. The information in the previous validation document had not been up-dated since 1988,[15] although some of the information on alternative approaches was previously contained in R6 Appendices. The validation consists in parts of brief summaries which provide a quick overview, and of detailed information for selected tests. This latter information also serves as worked examples of the procedures. In total there are four parts to Chapter V:

V.1   Validation of Basic Procedures
V.2   Validation of Alternative Approaches
V.3   Worked Examples of Basic Procedures
V.4   Worked Examples of Alternative Approaches

4. WORLDWIDE DEVELOPMENTS

In the last few years there has been a remarkable number of developments worldwide in the production or updating of guidelines for flaw evaluation. These include:

- the SINTAP procedures[2] described in Section 2;
- the R6 Revision 4[3] procedures described in Section 3;
- British Standards developments leading to BS7910;[5]
- the RSEM procedures[21] developed specifically for nuclear power plant applications in France;
- a flaw evaluation handbook produced in Japan in support of a Japan Society of Mechanical Engineers (JSME) fitness-for-purpose code;[22,23]
- a Chinese procedure being proposed as a Chinese national standard;[24]
- a comprehensive fitness-for-purpose guide, API 579,[6] published by the American Petroleum Institute.

All of these procedures have common features. Many use failure assessment diagrams. Others use reference stress methods to develop estimates of crack driving forces. All the procedures require common inputs such as a stress intensity factor solutions, limit load solutions, yield stress data and fracture toughness data. The SINTAP project has demonstrated that by building on these common features, significant progress can be made rapidly by collaborative effort.

Within Europe, a Comité Européen de Normalisation (CEN) working group has been formed under the Technical Committee 121 'Welding'. The working group has recommended the publication of BS7910,[5] with a European foreword, as a European technical report. The working group has also recommended TC121 to initiate activities to establish a number of European standards covering the application of methods for assessing the acceptability of flaws in metallic structures involving other key technical committees in joint working groups. The working group has recommended the establishment of a new work item for a European standard on methods for assessing imperfections in metallic structures based on BS7910, SINTAP and results of other ongoing research projects.

5. CONCLUDING REMAKS

Although fundamental concepts for flaw assessment, such as the failure assessment diagram, are well established in several codes the treatment of input data and the ability to maximise the value of such data through a hierarchical approach remain key development areas. In both SINTAP and R6 Revision 4, recent work on modified FADs to reflect actual materials' behaviour has been incorporated and enhanced methods for data treatment added. The structure of both procedures have also been rationalised in a hierarchical manner such that the advantages of increasing data quality in reducing conservatism are evident.

Worldwide developments also suggest an increasing activity in this field in recent years. However, within Europe development towards an EU standard will progress within CEN, comprising further development of the current UK standard BS7910 coupled with input from ongoing and completed research activities.

ACKNOWLEDGEMENTS

Financial support for the CEC for the SINTAP project performed under the EU contract CT95-0024 is gratefully acknowledged. This paper is published with permission of Corus Research, Development & Technology, and British Energy Generation Ltd.

REFERENCES

1. U. Zerbst, R A. Ainsworth and K.-H. Schwalbe, 'Basic Principles of Analytical Flaw Assessment Methods', *Int. J. Press. Ves. Piping*, 2000, **77** (14–15), pp. 855–867.
2. S. Webster and A. Bannister, 'Structural Integrity Assessment Procedure for Europe – of the SINTAP Programme Overview', *Engng. Fract. Mech.*, 2000, **67** (6), pp. 481–514.
3. R6 Revision 4, 'Assessment of the Integrity of Structures Containing Defects', British Energy Generation Ltd, Report R6 Revision 4, 2001.
4. H. G. Pisarski and K. Wallin, 'The SINTAP Fracture Toughness Estimation Procedure', *Engng. Fract. Mech.*, 2000, **67** (6), pp. 613–624.
5. British Standard BS7910 'Guidance on Methods for Assessing the Acceptability of Flaws in Metallic Structures', British Standards Institution, 2000.
6. T. L. Anderson and D. A. Osage, 'API 579: A Comprehensive Fitness-for-Service Guide', *Int. J. Press. Ves. Piping*, 2000, **77** (14–15), pp. 953–963.
7. K. H. Schwalbe, U. Zerbst, W. Brocks, A. Cornec, J. Heerens and H. Amstutz, *The Engineering Treatment Model for Assessing the Significance of Crack-like Defects in Engineering Structures (EFAM-ETM 95)*, GKSS Forschungzentrum, 1996.
8. V. Kumar, M. D. German and C. F. Shih, 'An Engineering Approach for Elastic-Plastic Fracture Analysis', General Electric Company, NP-1931, Research Project 1237-1, Topical Report, 1981.
9. I. Milne, R. A. Ainsworth, A. R. Dowling and A. T. Stewart, 'Assessment of the Integrity of Structures Containing Defects', *Int. J. Press. Ves. Piping,* 1988, **32**, pp. 3–104.
10. Y.-J. Kim, M. Koçak, R. A. Ainsworth and U. Zerbst, 'SINTAP Defect Assessment Procedure for Strength Mis-matched Structures', *Engng. Fract. Mech.*, 2000, **67** (6), pp. 529–546.
11. R. A. Ainsworth, A. C. Bannister and U. Zerbst, 'An Overview of the European Flaw Assessment Procedure SINTAP and its Validation', *Int. J. Press. Ves. Piping*, 2000, **77** (14–15), pp. 869–876.

12. P. J. Budden, J. K. Sharples and A. R. Dowling, 'The R6 Procedure: Recent Developments and Comparison with Alternative Approaches', *Int. J. Press. Ves. Piping*, 2000, **77** (14–15), pp. 895–903.
13. R. A. Ainsworth, J. K. Sharples and S. D. Smith, 'Effects of Residual Stresses on Fracture Behaviour – Experimental Results and Assessment Methods' *J Strain Analysis*, 2000, **35** (4), pp. 307–316.
14. R. A. Ainsworth, F. Gutierrez-Solana and J. Ruiz Ocejo, 'Analysis Levels Within the SINTAP Defect Assessment Procedures', *Engng. Fract. Mech.*, 2000, **67** (6), pp. 515–527.
15. I. Milne, R. A. Ainsworth, A. R. Dowling and A. T. Stewart, 'Background to and Validation of CEGB Report R/H/R6-Revision 3', *Int. J. Press. Ves. Piping*, 1988, **32**, pp. 105–196.
16. F. M. Burdekin, 'Report of TAGSI Task Group on Crack Arrest-2', TAGSI/P(98)157, 1999.
17. TAGSI Fracture Subgroup, 'Draft response to RRA/MoD questions on interactions between adjacent defects', TAGSI/FSG(95)P97(Draft 2), 1997.
18. A. G. Miller, 'Review of Limit Loads of Structures Containing Defects', *Int. J. Press. Ves. Piping*, 1988, **32**, pp. 197–327.
19. H. Tada, P. C. Paris and G. Irwin, *The Stress Analysis of Cracks Handbook*, third edn, ASME, 2000.
20. Y. Murakami, *Stress Intensity Factors Handbook*, vols 1, 2, Pergamon, 1987.
21. C. Faidy, 'RSE-M. A General Presentation of the French Codified Flaw Evaluation Procedure', *Int. J. Press. Ves. Piping*, 2000, **77** (14–15), pp. 919–927.
22. H. Kobayashi, S. Sakai, M. Asano, K. Miyazaki, T. Nagasaki and Y. Takahashi, 'Development of a Flaw Evaluation Handbook of the High Pressure Institute of Japan', *Int. J. Press. Ves. Piping*, 2000, **77** (14–15), pp. 929–936.
23. H. Kobayashi and K. Kashima, 'Overview of JSME Flaw Evaluation Code for Nuclear Power Plants' *Int. J. Press. Ves. Piping*, 2000, **77** (14–15), pp. 937–944.
24. P. N. Li, Y. Lei, Q. P. Zhang and X. R. Li, 'A Chinese Structural Integrity Assessment Procedure for Pressure Vessels Containing Defects', *Int. J. Press. Ves. Piping*, 2000, **77** (14–15), pp. 945–952.

CHAPTER 5

Fatigue-Based Assessments in Aircraft Structures – Designing and Retaining Aircraft Structural Integrity

P. E. Irving

Cranfield University, Cranfield, Beds MK43 0AL, UK

ABSTRACT

The processes of fatigue substantiation of civil transport aircraft are reviewed. The requirements for safe life, fail safe and damage tolerance design are described and related to the written requirements of JAA and FAA regulators. The substantiation of fixed wing aircraft is contrasted with requirements for rotary wing aircraft, where safe life design rather than damage tolerant is still the norm. The contribution of Health and Usage Monitoring Systems (HUMS) to safety in safe life helicopter designs is assessed. Challenges to the fail-safe/damage-tolerant approach posed by ageing aircraft, and the introduction of alternative materials such as carbon fibre polymer composites are described. The actions taken by regulators, manufacturers and operators to maintain airworthiness in the face of these challenges are described.

INTRODUCTION

Fatigue failure has been with the aircraft industry from the very beginning. Even before the Wright brothers' first flight in 1903, a fatigue failure in an engine shaft necessitated a postponement of the first attempt while a new shaft was manufactured. In the succeeding century, fatigue failures have been brought under control to the extent that fatal accidents resulting from failure of the aircraft structure occur approximately once every 100 million flights. This low rate arises not from technologies which have eliminated fatigue; conversely it has arisen from an acceptance that fatigue damage is inevitable, that aircraft structures contain flaws, and that the most accurately calculated lives may be in error.

In this paper procedures for design against fatigue and avoidance of failure in civil aircraft structures are reviewed. The challenges posed by ageing aircraft, and by the pressures to reduce manufacturing costs by use of new materials and processes are described and discussed.

69

CIVIL AIRCRAFT ACCIDENT RATES

Despite the growth in air transport passenger hours of about 6% per year, civil aircraft fatal accident numbers have remained roughly constant over the past 20 years at between 20–30 per year world wide.[1] As the number of flights has increased, the number of fatal accidents per flight has reduced in the same period, from 1 every 10 million flights in 1980 to about 1 every 30 million flights today. As Fig. 1 shows, there is considerable year-to-year variation in the rates, but the trend is down. The majority of these accidents are due to human error and environmental factors. Failures of the aircraft or engine structure are a minority of the total. A recent survey of aircraft accidents in the UK showed that only 10% of recent accidents could be attributed to airworthiness causes.[2] However, there are 10 non fatal accidents for every fatal one, and additionally there are many fatigue failures which do not result in serious accidents. This represents a problem with economic rather than human dimensions. What is the cost of detection, repair and maintenance necessary to avoid fatigue and corrosion failures becoming fatal accidents?

As flights and flight hours continue to increase in the 21st century, public perception of the safety of flight as a form of transport will depend on fatal accident numbers continuing to stay constant. Fatal accident rates, whether due to airworthiness causes or human error, must continue to decline. Research is underway to improve the rates for both human factor and airworthiness related accidents. For airworthiness accidents, the challenges lie in two areas.

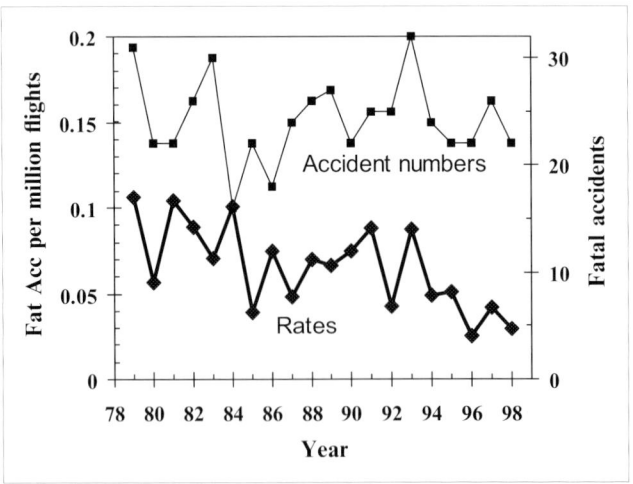

Fig. 1   World fatal accident numbers and fatal accident rates 1978–98. Large Civil Jet Transport aircraft Data from Ref. 1.

Firstly, the continuing presence in the world aircraft fleet of 4000–5000 aircraft of calendar age greater than 15 years, is of potential airworthiness concern. Many of these aircraft have accumulated flights and flight hours approaching and in many cases exceeding the original aircraft design goal. The challenge is to ensure that degradation of these ageing aircraft due to corrosion and fatigue-induced multi-site and multi-element damage does not cause age-related failures and accidents.

Secondly, the continuing introduction of new aircraft designs, containing new materials such as polymer and metallic composites and fabricated by cost-efficient processes such as casting, welding and adhesive bonding, must take place without any unexpected failures leading to airworthiness accidents.

DESIGNING AGAINST FATIGUE IN AIRCRAFT

Despite the fact that fatigue undoubtedly caused fatal aircraft accidents,[3] for the first 50 years of the aircraft industry, there was little emphasis on design against fatigue. Propellers, particularly aluminium propellers had been life limited since the 1920s, but this was done without recognition of the role which usage changes might play in determining the overall life. Bearings similarly were life limited by use of the Palmgren rule.[4]

It was not until 1943 that Bland and Sandorff [5] suggested the use of finite life designs for aircraft structures, noting that attempts to design aircraft structures as infinite life had been unsuccessful as evidenced by the number of fatigue failures which occurred. Furthermore as evolution of aircraft design continued, wing loadings were increasing. From 1935–1942 wing loadings increased from 18 lbs/sq ft to 28 lbs/sq ft,[5] ($88 \, \mathrm{kg \, m^{-2}}$ to $137 \, \mathrm{kg \, m^{-2}}$), and have continued their increase until the present day, with the Boeing 747 at 140 lbs/sq ft ($685 \, \mathrm{kg \, m^{-2}}$). Infinite life design, with the emphasis on keeping the stresses less than the endurance limit was going to result in very heavy aircraft, and the poor previous history of failed infinite life designs was not encouraging.

Bland and Sandorff suggested a finite life design approach in which inputs from measured service loads and constant amplitude fatigue properties could be combined to assess the life limit of the structures. The approach suggested was formalised by Miner[6] in his 1945 paper on what has become known as Miner's linear damage rule, or the Palmgren–Miner rule. This was to result in lighter structures with greater freedom from fatigue failure. This approach is the forerunner of the so called Safe Life approach to design against fatigue.

One of the first aircraft to be designed to a life limit was the Comet aircraft.[7] The Comet accidents of January and April 1954, in which the aircraft disintegrated at cruising altitude of 35 000 feet off the coast of Italy, provide a fine example of why fixed life approaches to design against fatigue are in themselves insufficient. No matter how accurate and sophisticated the prediction procedures, there will always be some probability of the calculated life being in excess of the service life and a catastrophic fatal accident resulting. A design approach which goes beyond the safe life calculations is required.

The safe-life of an aircraft has had many definitions and specifications over the last 50 years. There appears to be no generally agreed consistent definition.[8] Typical is the one given by the FAA in 1998:

> the number of events such as flights, landings or flight hours, during which there is a low probability that the strength will degrade below its design ultimate value due to fatigue cracking.

In fixed wing aircraft in civil aviation, the safe life approach is now confined to the under-carriage and the engine assemblies. All other parts of the aircraft must now be designed using the damage tolerant approach. In military aircraft, variants on the safe life approach are still in use (for example in the British RAF and the US Navy). The US air force and the European, Canadian and Australian Airforces all use the damage tolerant approach, although its use will differ in detail from procedures implemented by the FAA (US) and JAA (European, including the UK CAA) civil aircraft regulators. Rules governing civil aircraft design and operation are almost completely harmonised between JAA and FAA organisations.

In the case of the Comet aircraft,[7] detailed investigations coupled with full-scale testing after the accidents revealed that stress values of the fatigue cycles applied to the pressured cabin were greater than envisaged during aircraft design, resulting in a shortened fatigue life (less than 1000 ground-air-ground cycles for the aircraft which crashed.)

Swift[9] has demonstrated that in addition to the Comet aircraft calculated safe life being in error, there was the further problem that cracks which developed in the fuselage would have become unstable at relatively short crack lengths, thus reducing the probability of crack detection prior to catastrophic failure. Additional design procedures were required which covered the eventuality that if a non conservative error is made in safe life calculation, the aircraft would remain safe by detection of cracking before it reached a size where unstable fracture resulted. This approach became known as fail safe design and was recognised by regulators as important in amendments to the regulations after the Comet investigations.

In 1956 the British Civil Airworthiness Requirements (BCAR) were amended to include fatigue for the first time. The new sections[10] included the following:

the pressure cabin and local structure shall have satisfactory fatigue characteristics which shall be confirmed by an appropriate programme of tests. The tests shall be used either:

(a) to show that all possible failures are of a 'Safe' variety – i.e. the type and rate of propagation of cracks are such that they would be noted in a normal inspection before they introduced any likelihood of catastrophic failure;

or

(b) if (a) cannot be shown, to establish a 'Safe' life for the pressure cabin as a whole or its component parts.

These requirements, initially confined to the pressure cabin, were extended to the whole aircraft in 1959. The US-based FAA regulators also first addressed the fatigue issue in 1956, with a similar choice between safe-life or fail-safe approaches.

It is interesting to note that the fail safe requirements were imposed without the benefit of fracture mechanics models for fatigue crack growth, which were first published in 1961,[11] and remained contentious as a reliable means of calculating crack growth rates and lives for some years after this date.

Fail safety is therefore an approach to fatigue design which requires that an unforeseen crack will be detected before catastrophic failure. It infers use of materials with slow rates of fatigue crack growth and with high fracture toughness. It also implies use of multiple load path construction so that the failure of a single load path is not necessarily catastrophic. Fail-safe also uses crack stoppers: design features which will arrest or divert a running crack. The use of any or all of these aspects to protect an aircraft from catastrophic failure relies in the end on the crack being detected prior to final failure. Crack detection via a planned schedule of inspections is an essential part of the concept.

Derivation of inspection intervals without fracture mechanics models can only be based on component or structural testing, rather than calculation, and relies on the test set up and loads being representative of service. Estimation of probabilities of detection of cracks is similarly difficult to quantify. In civil aircraft today (and for the last 20 years), the fail-safe approach is strongly encouraged, but only within a damage-tolerant (quantitative crack growth) framework. It is not permitted on its own.[12] Military aircraft (e.g. USAF) have retained the option of a fail safe approach without damage tolerance via slow crack growth, within a strictly defined inspection regime.[13]

Since 1978[12] large civil transport aircraft structures have been required to be damage tolerant according to both FAA and JAA regulations. The only remaining parts of the aircraft permitted to use the safe life approach are the aircraft under-carriage and the engines. Although the regulations in theory always permit safe life as an alternative if damage tolerance cannot be demonstrated, in practice it has been found possible to demonstrate damage tolerance for all components except the under-carriage. Even in the case of the engines, components invariably have a safe life calculated using a flaw growth approach. Only the under-carriage remains with an initiation based safe life approach.

The regulations themselves (JAR 25.571 – Damage tolerance and fatigue evaluation of structure for large aeroplanes, 1998, and the equivalent FAA regulation) are coy on the definition of damage tolerance. In general terms they state that 'an evaluation of the strength, detail design and fabrication must show that catastrophic failure due to fatigue, corrosion or accidental damage will be avoided through the operational life of the aeroplane.'

In more detail, the regulations stipulate that the evaluations must include:

Typical loading spectra,

Identification of principal structural elements, failure of which could cause catastrophic failure of the aeroplane,

An analysis of the above elements supported by test evidence,

Service history of similar aircraft may be used in the evaluations,

Based on the evaluation, inspections or other procedures must be established as necessary to prevent catastrophic failure,

For a damage tolerant evaluation; this must include a determination of the probable locations and modes of damage due to fatigue, corrosion or accidental damage,

Analysis and test evidence is required in the evaluation,

The extent of damage for residual strength evaluation at any time within the operational life must be consistent with the initial detectability and subsequent growth [of the damage] under repeated loads,

The residual strength evaluation must show that the remaining structure is able to withstand static ultimate loads (specified as limit loads for various flight conditions)

There is a further requirement that the aircraft must be capable of completing a flight during which structural damage occurs as a result of bird impact, sudden decompression due to, for example, uncontained engine burst.

Confusingly, both the requirement to resist discrete damage and that to resist slow crack growth are referred to as damage tolerance. The regulations are supported by advisory circulars (ACJs) which provide suggested means by which compliance may be achieved; it is emphasised in these documents that the suggested means are not the only ways to achieve damage tolerant design.

Figure 2 shows degradation in aircraft strength with increasing fatigue crack growth, mechanical damage or corrosion. The end of aircraft life is the point at which the degradation is such that the aircraft cannot withstand limit load conditions – the largest loads encountered in the aircraft life. This is easy to illustrate schematically; in practice there are a range of specified limit load conditions which will be relevant to different parts of the aircraft structure.

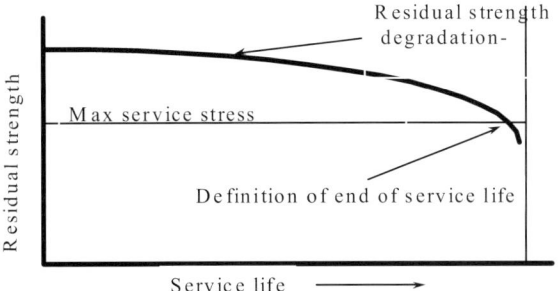

Fig. 2 Degradation in residual strength with time in service.

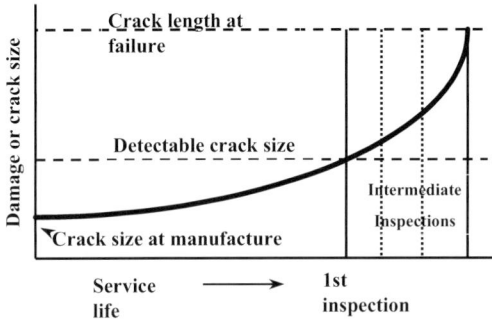

Fig. 3 Schematic illustration of damage growth and inspection intervals in damage tolerant design.

Figure 3 shows schematically the converse behaviour, the growth of structural damage from an initial defect size on entering service to the size at which limit load conditions cannot be met. Initial defects and initial growth will not necessarily be detectable using selected NDT techniques. At some point in the life, cracks and defects will be detectable with 90% probability and 95% confidence. This point, the inspection threshold, N_{th} is the service life at which it first becomes worthwhile inspecting the aircraft for cracks. The remaining life between the inspection threshold and life to failure, is the 'inspectable' region, and may be typically divided into two or three equal service life intervals with inspections at the end of each interval.

NDT CAPABILITY

The detectability of the damage is all important; without damage detection, as for fail-safe, there is no safety net. Analytical prediction of the growth of an assumed worst case defect is not sufficient. Defects must be detectable prior to catastrophic failure. The levels of detectability for calculation of inspection threshold and for the inspections must be high. 90% probability of detection, (POD) and 95% reliability are used. In principle any NDT technique may be used provided that the defect sizes used in the calculation are at the 90/95 levels for that technique. However, for the particular combination of materials and stress levels in a design, the selected technique must result in a high probability of damage detection within workable inspection intervals.

Rigorous validation of NDT techniques to determine the crack sizes for the 90/95 levels of detection has rarely been performed. The requirements on sample numbers and number of NDT operators, imposed by the statistical requirements for the experiment are formidable. See for example the work of Rummel.[14]

From the aircraft operators' viewpoint, it is economically desirable that life to inspection threshold should be as long as possible, provided that the remaining inspections do not occur so frequently as to pose operational problems. An NDT technique which can detect only relatively large crack sizes at the 90/95 level – such as visual inspection, will move the inspection threshold, Nth, to longer lives (Fig. 3). This implies that the fraction of total life where defects are inspectable becomes small. Reduction of initial manufacturing defect size will produce a similar effect. An aircraft life where the region of detectable crack growth is less than half of the total operational life may be unacceptable to regulators, as the uninspectable period prior to Nth will then form the majority of the aircraft life. Typically Nth is selected to be about a third to a fifth of the total life of the aircraft.

MATERIAL PARAMETERS

Fatigue crack growth and fracture toughness properties of materials play a major role in damage tolerance capability. For many aircraft situations, plane strain conditions will not be met at failure, and mixed mode conditions relevant to failure of thin sheet will occur. Conditions for unstable propagation of cracks in fuselage or wing structures are complex and difficult to calculate. Details of the experiments and predictive models used will not be described here – they are often conducted in support of certification programmes and are highly confidential. See however examples of residual strength calculations in Refs 9 and 15.

As Fig. 3 shows, the crack length at failure does not influence the overall life significantly, but it will influence the probability of detection of cracks prior to catastrophic failure. Figure 4, showing the relation between crack length and probability of detection (POD) illustrates the dramatic effect that crack size has on POD for a defined inspection technique. Swift[16] gives a number of examples of

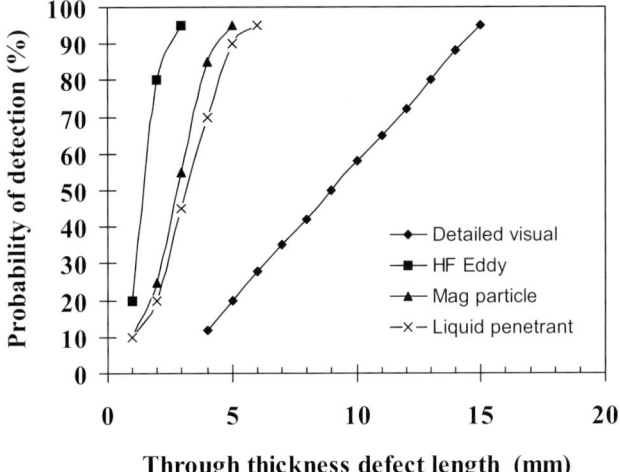

Fig. 4 Defect size vs. probability of detection for different NDT techniques at 95% confidence.

how capability to resist the presence of cracks in excess of 500 mm long, has allowed unexpected cracks to be detected prior to a catastrophic event.

Figure 5 shows a comparison of 2024, 7010 and 7075 aluminium alloys in terms of their fatigue crack growth resistance. Figure 6 compares crack growth rates in 2024 aluminium with those in titanium and steel. The well-established superiority in fatigue crack growth resistance of 2024 over 7XXX aluminium alloys is shown.

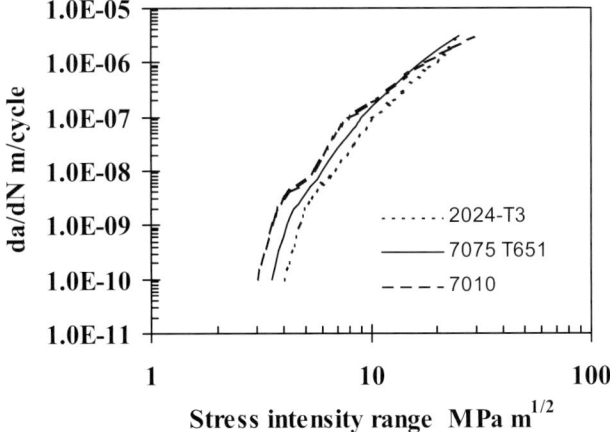

Fig. 5 da/dN vs. ΔK for aluminium alloys showing superiority of 2xxx over 7xxx alloys at $R = 0.1$.[17]

Similarly, the superiority of titanium over aluminium alloys, and the superiority of steels over titanium, is demonstrated in Fig. 6, arising because of the modulus differences in the three alloys. For damage-tolerant design, the comparison of growth rates is more complex than the ΔK based comparison of Figs 5 and 6, as the different strengths of each alloy will change the design stresses used in each one, and this will change the ΔK value of cracks of a specified length. This can be illustrated by integrating the Paris Law,[11] to obtain an expression for cycles to grow cracks between a start and finishing crack length:

$$\mathrm{d}a/\mathrm{d}N = C(\Delta K)^m$$

expresses crack growth rates in terms of stress intensity range ΔK and the constants C and m.

ΔK is given by:

$$\Delta K = \Delta\sigma\sqrt{\pi a}\beta$$

Where $\Delta\sigma$ is the nominal stress range, a the crack length and β a factor dependent on component geometry and crack length.

Integrating to calculate fatigue cycles to grow a crack from initial length a_o to final length a_f gives the a relation of the form:

$$N = \frac{1}{\pi^{m/2}C\Delta\sigma^m}\left(a_0^{(1-m/2)} - a_f^{(1-m/2)}\right)$$

Ignoring β, the geometry correction factor, for simplicity.

This expression shows that in addition to the starting crack length a_o, and the material constants C and m, life is inversely dependent on stress range $\Delta\sigma^m$ As m

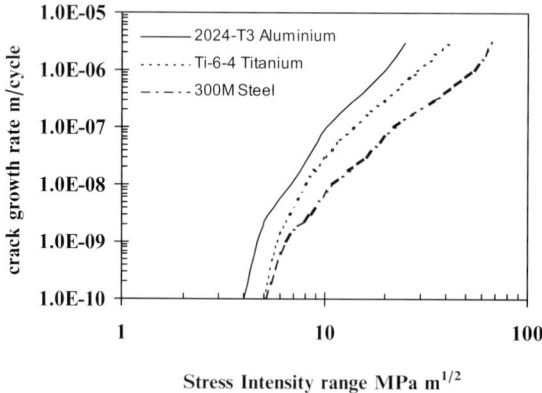

Fig. 6   Comparison of fatigue crack growth properties for steel, titanium and aluminium at $R = 0.1$.[17]

is typically between 2 and 4, this will have a powerful effect on life. Materials of different static strengths used at stresses of the same fraction of their strengths, will have very different fatigue crack growth lives. This dependence of life on stress range, rather than poor fatigue crack growth resistance, is responsible for the well appreciated lack of damage tolerance in structures manufactured from high-strength materials.

More quantitative comparison of the damage tolerance capability of different materials under real service load spectra and real geometries may be performed by using a computer-based crack growth integration package.[18] This can perform numerical integration of crack growth rate data for a single component geometry, defined start and finish crack lengths, and a realistic service load spectrum. Using this approach, materials of different strengths can be compared on the basis of lives produced for a constant ratio of maximum spectrum stress to proof strength. Figure 7 shows schematically the approach.

To produce the data shown in Figs 8 and 9, a single edge notch sample was used subjected to the TWIST transport aircraft wing loading spectrum, with the largest stress excursion in the spectrum being set equal to 60% of the static 0.2% proof strength of each alloy. The number of individual load cycles to grow a crack between 1 mm and 13 mm was calculated using the widely available crack growth calculation package AFGROW.[18] Three aluminium alloys, 7075, 2024 and 6061, are compared in Fig. 8 and aluminium 2024, titanium 10–2–3 and A533 steel are compared in Fig. 9.

For aluminium alloys, longest lives are obtained for the lowest strength alloy, 6061. The additional strength of the 7075 alloys can only be exploited in damage-tolerant design, at the expense of reduced life, as greater design stresses will

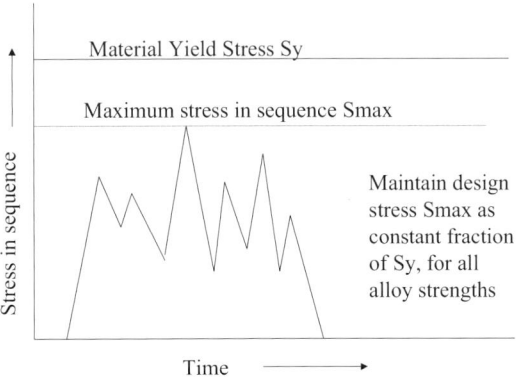

Fig. 7   Maximum stress in spectrum is defined as design stress; design stress should be a constant fraction of yield strength for all strengths of alloys.

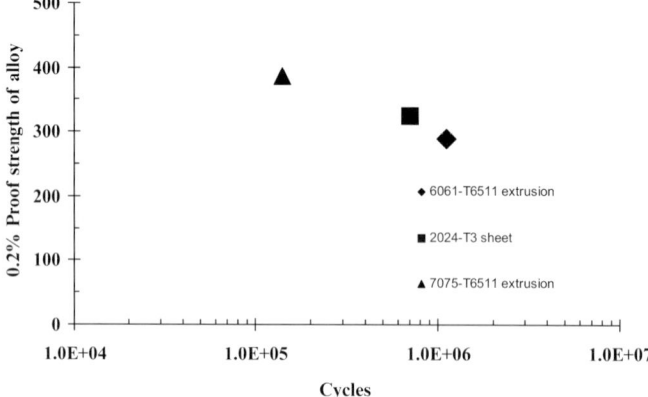

Fig. 8   Plot of cycles to grow crack between 1 and 13 mm in SEN tension sample subjected to TWIST loading for three different aluminium alloys. Largest stress in TWIST sequence is 60% of 0.2% proof strength for each alloy.

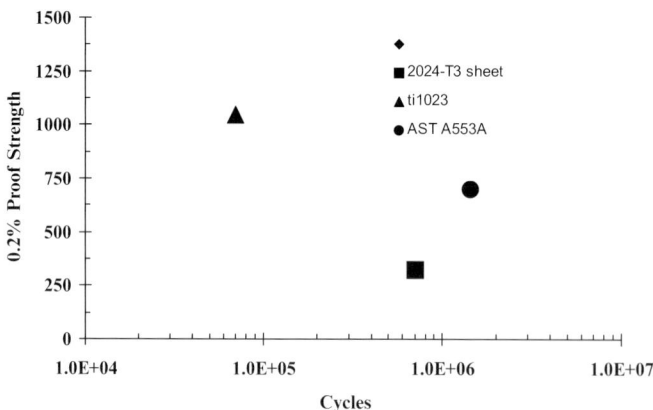

Fig. 9   Plot of cycles to grow crack between 1 and 13 mm in SEN tension sample subjected to TWIST loading for 2024 aluminium, titanium 10–2–3 and A553 A Steel. Largest stress in TWIST sequence is 60% of 0.2% proof strength for each alloy.

produce shorter lives (Fig. 8). If it is required to produce the same life as 2024 T3, the stress levels must be reduced, using less of the 7075 static strength. Figure 9 shows results of a similar calculation comparing 2024 T3, titanium 10–2–3, and a pressure vessel steel A533. Lowest strengths again give longer lives, although the superior toughness and stiffness of the A533 steel gives it additional life over the 2024, even though the latter has the lower strength.

The reduced fatigue life of 7XXX aluminium alloys in damage-tolerant design, means that their use on large passenger aircraft is confined to regions of largely compressive stress such as the upper wing structures, where their increased strength can be exploited to reduce wing weight without the reduced life which would result if they were applied in tension regions of the structures such as lower wing and fuselage. In military aircraft the greater value placed on reduced weight at shorter lives in damage-tolerant designs results in widespread use of 7XXX alloys all over the aircraft, together with titanium where its greater strength over aluminium can be exploited.

DAMAGE TOLERANCE vs. SAFE LIFE DESIGNS

Components such as undercarriages are manufactured of ultra high strength steel such as 300M, with design stresses correspondingly high. For the reasons outlined above, the fraction of the 300M strength usable in damage-tolerant design will be small. Hence these components are designed using the safe life approach – essentially an 'initiation' based method. Fatigue initiation behaviour is strength level dependent, and high strength steels will have superior fatigue crack initiation strength to that of low strength steels, and will be better able to resist the higher design stresses. Use of safe life design allows the tremendous strength to weight properties of 300M to be exploited to reduce undercarriage weight. The relative insensitivity of cracked components' fatigue life to strength improvements is illustrated by a comparison of Figs 10(a) and (b).

Figure 10(a) shows a comparison of a *S–N* curve for a smooth specimen of 2024 T3 compared with the calculated *S–N* curve for the same sample with a 1 mm surface crack. Figure 10(b) shows the same comparison for a 4340 quenched and tempered steel of similar strength to 300M. The difference between the cracked and uncracked samples is greater for the higher strength material, (424 MPa as opposed to 165 MPa for the aluminium). Both materials experience similar percentage reductions in strength – about 80% in each case.

DAMAGE TOLERANCE IN HELICOPTERS

Helicopter components, despite the encouragement of regulators, are still largely designed using the safe life approach.[20] The reason for this is their use of high strength steel and titanium alloys in the transmission, gearbox and rotor control systems. Fatigue substantiation has traditionally been performed using safe life which fully exploits the high strength of these alloys to resist fatigue initiation. Transference of these safe life stress levels to a damage-tolerant design without

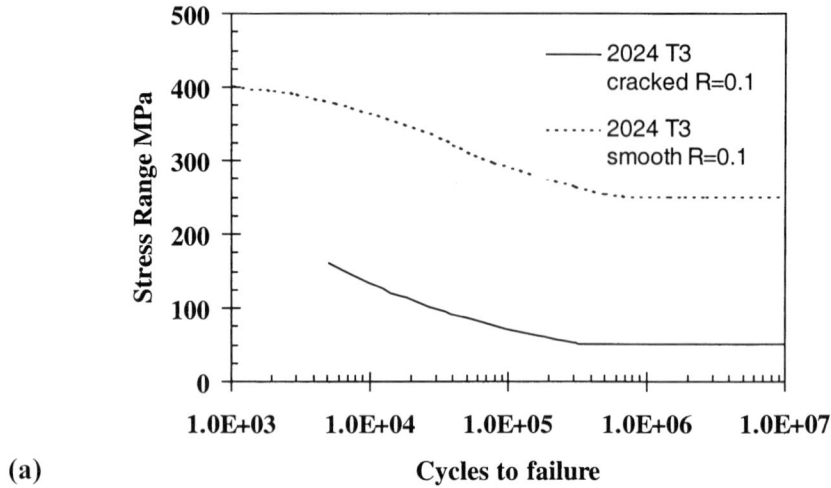

(a)

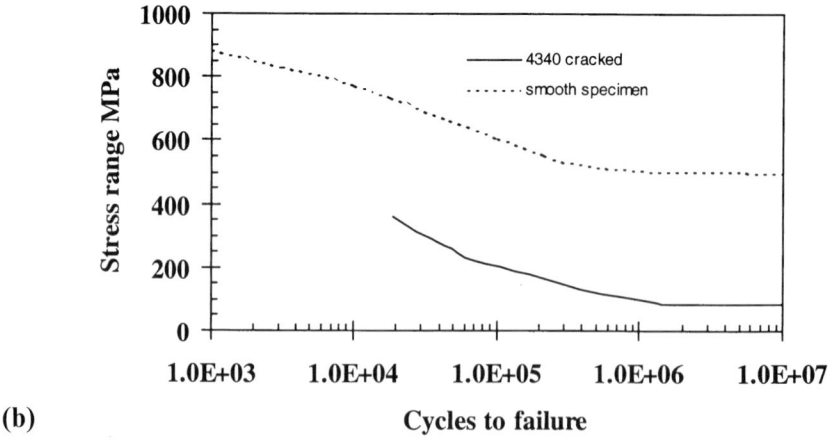

(b)

Fig. 10(a)   Cracked and uncracked *S–N* curves of 2024 T3 showing the degradation which a 1 mm crack causes. Smooth specimen data from Ref. 19, corrected from $R = -1$ to $R = 0.1$; crack growth data for calculation of cracked endurance curve is from Ref. 17. (b) Comparison of cracked and uncracked 4340 quenched and tempered steel (0.2% proof 1400 MPa), showing degradation caused by 1 mm crack.[18, 19]

change of stress level or material means that the fatigue crack growth cycles are few, and demonstration of damage tolerance becomes correspondingly difficult.

There is the further factor that helicopter load spectra contain rotor induced load cycles at either rotor frequency – about 5 Hz – or blade passing frequency – about 20 Hz. There are other multiples of these frequencies in the spectrum but their load ranges are small in comparison with the first two. These are superimposed on mean stresses which change with frequencies of 1–0.1 Hz. These changes result from normal helicopter manoeuvres. In a safe life design, the stress range of cycles resulting from rotor motion are kept less than the endurance limit of the alloy concerned, and are hence undamaging. The cycles arising from manoeuvres alone are used in the damage analysis. As Fig. 10 shows, in a damage tolerant design, stress cycles which will not cause crack advance for a sample with a 1 mm crack need to be very much smaller than cycles which are 'undamaging' in a fatigue initiation calculation.

All of these factors mean that in helicopters, the window of crack growth in which the crack is detectable but has not yet attained a critical crack size for failure, is a small fraction of total life, and inspection intervals have been calculated[21] to be as small as a few hours. Under these circumstances operation of a damage-tolerant design becomes uneconomic.

The consequence is that inadvertent events which curtail the designated safe life render the helicopter vulnerable to sudden airworthiness failure. The accident to a Chinook helicopter (Fig. 11) in 1986 off Shetland in the North Sea is typical. The accident originated in the failure of a synchronisation gear in the transmission system, shown in Fig. 12, which allowed the fore and aft rotors to collide. The helicopter subsequently plunged into the sea with loss of 43 lives. The failure originated in a design modification made to the gear assembly, causing an eventual fretting fatigue failure. In the absence of an adequate inspection window to detect the progress of the failure, disintegration of the gear was unexpected and proceeded with such rapidity that no emergency landing was possible.

The lack of damage tolerance in helicopters has been ameliorated in recent years, with the development of Health and Usage Monitoring Systems (HUMS) for helicopter transmission systems. These devices use computer-based vibration diagnostic techniques[22] to assess wear and fatigue crack development in transmission system components. The advantage of the systems is that they continuously monitor the components for failure, effectively reducing the inspection interval to zero. Provided that the life remaining after defect detection is sufficient to allow the helicopter to land safely, HUMS systems appear a safe substitute for damage tolerance.

The major current disadvantage is that the sensitivity of the detection techniques at present is such that if damage is detected, the helicopter is required to land immediately with some types of warning indication, as there may be little

Fig. 11   Chinook helicopter.

remaining life after detection of a fault. Thus HUMS diagnostics are tools for a 'retirement for cause' approach to damage tolerance. Figure 13 shows schematically how HUMS systems operate within a damage tolerance context.

Other disadvantages of the current generation of vibration diagnostic systems include the fact that there is insufficient discrimination between failed and non-failed conditions. There is therefore a substantial incidence of false alarms. In addition the devices require 'tuning' to specific aircraft in order to set the critical levels in the monitored parameter, exceeding which will trigger an alert.

Figures compiled by CAA[23] demonstrate that over the first five years of HUMS operation on heavy civil helicopters operating in the North sea, HUMS operation was highly successful. There were 160 aircraft reviewed, accumulating in excess of 0.5 million flying hours. Some 72 arisings (events of potential airworthiness concern which had led to significant maintenance action) had been recorded up to 1997. Of these 63 were airworthiness related. In 9 of these, the HUMS failed to

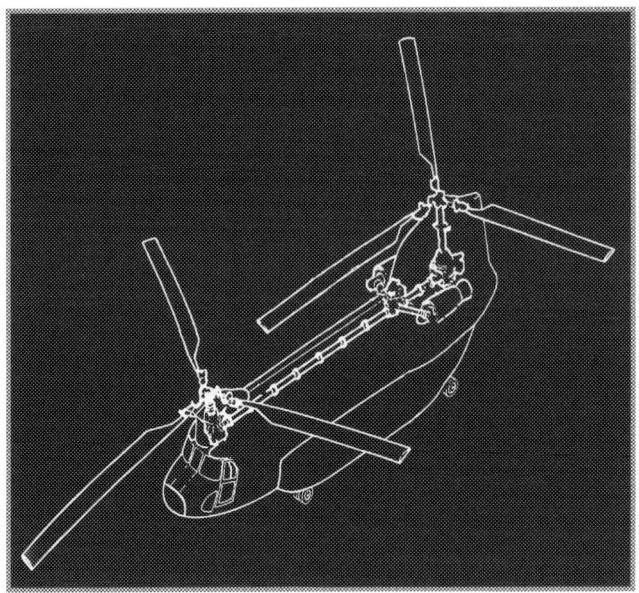

Fig. 12 Schematic diagram of Chinook transmission system.

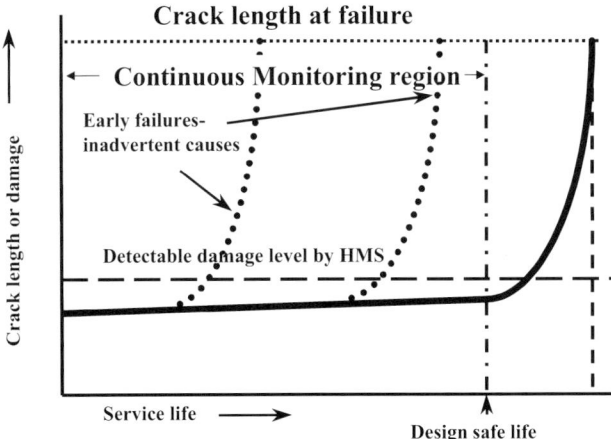

Fig. 13 Schematic illustration of the role of health monitoring systems in detecting inadvertent failures occurring at less than design safe life.

indicate a fault. In 43 arisings, the HUMS correctly indicated a fault, and in 9 of these, the fault was evident in the data but was not acted upon in a timely and correct manner. Six of the 63 were identified as potentially catastrophic, and would probably have caused an accident, had they remained undetected. The level of correctly identified faults and the probability that lives and significant costs have been saved encourages the belief that HUMS and related systems constitute the way forward for damage detection in aircraft.

Despite the title acronym HUMS, current generations of the devices are limited to health monitoring only. The addition of usage monitoring capability i.e. measurement of the loading environment, would allow development of prognosis capability in addition to detection and diagnosis capabilities. Prognosis would allow accurate assessment of the future probability of failure of components. Usage monitoring systems are valuable in permitting actual calculation of the fatigue damage levels received by a component, as opposed to the damage implied by the conservative loadings assumed in fatigue certification. Usage monitoring may thus allow life extensions or maintenance credits in a safe life design, and could also enable a damage tolerant fatigue substantiation. Use of real load data in fracture mechanics calculations could transform the very short crack growth lives obtained using conservative design loads, into much longer lives resulting from use of smaller stresses. The increased lives may prove adequate to demonstrate damage tolerance with economically viable inspection intervals.[24]

In practical terms there is little real difference between the traditional damage tolerance methodology of three inspections at intervals, and one of a retirement or repair as soon as a fault is detected. This is because when a fault is detected on a fixed wing aircraft inspection, the aircraft is not put back into service and flown with the crack being monitored until it achieves a sufficiently large size. It is repaired immediately. In the case of non-repairable items such as gears or bearings, they are replaced. There is thus no practical distinction between damage tolerance and a retirement for cause approach applied via a HUMS system.

In the future HUMS devices may achieve sufficient sensitivity and reliability that damage may be detected, growth monitored, and future life predicted with high levels of confidence over the entire life of the component. When this stage of development occurs, current criteria on when components and structures should be repaired or replaced will need to be drastically revised.

FLAW TOLERANT SAFE LIFE

Regulators have recently introduced an approach to fatigue substantiation of helicopters intermediate between safe life and damage tolerance. The flaw-tolerant safe life approach is offered as an alternative to both safe life and damage tolerance approaches.[18] In concept it is similar to safe life, except that 'non pristine' samples are used to determine the S–N behaviour of the material. 'Non pristine' samples are ones containing defects such as dents, scratches and corrosion pitting which might arise during service of the component. Having established the S–N data, the safe life is calculated using the material data and the service load data in the same way as for the normal safe life calculation. The S–N curve will be degraded from the S–N curve obtained on pristine samples, and will be intermediate between this (a best case) and the S–N curve measured on a pre-cracked sample (a worst case).

The approach has yet to be used for fatigue substantiation of a helicopter, and is currently the subject of considerable discussion between regulators and manufacturers regarding the practical details of its implementation. At worst the defects can be crack like, and an S–N curve approximating to that of the cracked components could be produced. At best the S–N curve of the non pristine components would approach the strength of the pristine samples, if benign defects were used. The approach does not provide the safety net which a damage-tolerant approach would do. However if it reduces design stress values, it could make the calculated safe lives more conservative, and because of the lower stress levels crack growth rates would be reduced. Inspection intervals could be increased under the reduced stress levels. The reduced stress levels may impose a weight penalty which manufacturers may not find acceptable.

AGEING AIRCRAFT ISSUES

According to a survey conducted in 2000,[25] there are around 5000 transport aircraft in the world between 15 and 25 years old. The definition of an ageing aircraft is one over 15 years old. A small number of aircraft types comprise much of the fleet – 737 (860 aircraft); 747 (480 aircraft); A300 (213 aircraft); A310 (74 aircraft) . One of the difficulties of monitoring the age of the aircraft[8, 26] is that age may be based on three counting concepts, none of them quantitatively related to fatigue or corrosion damage, but all of them approximately so. Age may be calendar time, number of flights, or flight hours. An aircraft may have design service goals set by its manufacturer, which can be expressed as, for example for the Boeing 737, as 75 000 flights, 51 000 hours, or 20 calendar years.

Table 1 shows fleet average, fleet leader and design goal for some common aircraft types, illustrating both the discrepancies between the different measures of age and the differences between the fleet leader lives and the fleet average.

In service, under the damage tolerance approach, there is no natural end point to the aircraft life, unless it is imposed separately. Aircraft may be inspected, repaired if necessary and flown until the inspection interval has been completed, inspected

Table 1   Selected aircraft types over 15 years old as at June 2000.[17]

Aircraft type	Total in fleet	Total over 15 years	Design life flights	Design life hours	Design life calendar years
A300	**490**	**213**	**36 000**	**60 000**	**20**
Fleet avrg			15 200	27 200	13
Fleet leader			35 405	55 174	25
A310	**255**	**74**	**35 000**	**60 000**	**20**
Fleet avrg			11 600	29 800	12
Fleet leader			25 681	58 682	17
747-100/200/300	**934**	**480**	**20 000**	**60 000**	**20**
Fleet avrg			10 000	46 800	14
Fleet leader			34 531	114 823	32
737-100/200	**1144**	**860**	**75 000**	**51 000**	**20**
Fleet avrg			42 900	46 900	22
Fleet leader			96 528	88 457	33

and flown ad infinitum. It is only since the 1998 revision of section 25.571 by the FAA,[27] that a design service goal has had any regulatory significance. Prior to this amendment, still (as of 2001) not agreed by the JAA, design lives were an internal figure used by manufacturers and had no other significance.

Another factor which exacerbated the effects of increasing aircraft age was the fact that many aircraft operating in the 1980s and 1990s were to designs certified prior to 1978 – the date at which full regulatory implementation of damage-tolerant design philosophy was made. Many of these aircraft did not have inspection intervals calculated using fracture mechanics. Often, as no problems were reported on the inspections, the intervals before the next inspection were extended, thus reducing the possibility of damage detection.

Ageing aircraft damage highlights one of the major, often hidden, assumptions of damage tolerance based fatigue substantiation, that the damage and fatigue crack shall be single source. The original intention of the damage tolerance approach was to design against the single rogue flaw, or discrete accidental damage. The fatigue crack growth techniques described earlier do this very successfully. With increasing aircraft age, corrosion and fatigue damage can occur at multiple sites over the aircraft structure. This is not in itself a problem if the sites are spatially well separated. The danger is that distributed damage or fatigue cracks, which are individually too small to be reliably detected, may collectively act to reduce the residual strength of the structure below that required to resist design limit load.

That this is a real danger is clearly demonstrated by the failure in flight of large sections of the upper fuselage structure of a Boeing 737 aircraft belonging to

Aloha Airlines in 1988 (Fig. 14). Amazingly, this catastrophic failure did not cause complete disintegration of the aircraft structure, and the remainder of the aircraft landed safely with only one fatality – perhaps a tribute to the inherent damage tolerance of the aircraft fuselage. Investigators subsequently attributed the failure to widespread fatigue damage (WFD – a form of damage where large parts of the structure have distributed multiple site damage (MSD) where large numbers of the rivet holes had small fatigue cracks emanating from them). The aircraft in question had been inspected for this form of damage in the months before the accident, without significant indications being reported. This confirms the difficulty of reliable detection of arrays of small cracks at riveted joints.

Ageing aircraft degradation had been recognised as a problem 10 years earlier in the late 1970s, when two accidents[28, 29] caused by structural failure, had occurred in aircraft despite their being designed to failsafe principles. Supplementary structural inspection programmes (SSIP) had been set up at that time to examine a range of identified aircraft types, establish their condition, repair the aircraft and redefine the damage tolerance analysis and inspection intervals and procedures. Additional corrosion control programmes were also implemented at that time. These measures were mandatory in the UK only and applied only to a relatively small fraction of the aircraft at risk. In the rest of the world, most

Fig. 14 Aloha Airlines accident in 1988, probably caused by undetected multiple site damage and widespread fatigue damage.

notably in the USA, the measures were advisory and consequently were only partially implemented.

The Aloha 737 accident caused an explosion of activity in regulators, manufacturers and aircraft operators. The level of awareness of the potential seriousness of the problem had been raised dramatically. In the wake of a meeting held between regulators, manufacturers and operators in the weeks after the accident, urgent action was taken to fully implement the measures defined earlier, to raise the importance of detection and repair of MSD in aircraft, and to establish research programmes into all aspects of ageing aircraft degradation.

Research in the period since then has exhaustively studied the degradation processes of corrosion and fatigue in aged aircraft. Much of this work has been published in many conferences and journals, for example.[30] For the purposes of the present paper, the degradation processes may be summarised as follows.

Corrosion and fatigue act together in degrading aircraft structural strength. In the early part of aircraft life, corrosion protection schemes inhibit corrosion processes, and fatigue acts alone. At lives in excess of 15 years, corrosion damage begins to be noticeable. Many aluminium alloys are prone to corrosion pitting and development of exfoliation cracks. Both exfoliation cracks and corrosion pits can act as initiation nuclei, drastically reducing the time or cycles taken to develop fatigue cracks. Fatigue lives may be reduced by up to 60–70%[31] in the presence of corrosion pits of only 50 μm depth.

Possibly the most serious form of corrosion degradation is debonding of the sealing medium within lap joints. The sealer function is both to prevent moisture ingress and also to transfer a proportion of the joint load in shear between the panels, rather than directly at the rivet. Without the sealing, moisture enters the joint, local stresses at the rivet will be enhanced and local fatigue crack development on either side of the rivet hole, is accelerated. Corrosion-induced pitting within the hole, arising from moisture ingress, will aid this process. The process can occur over extended lengths of joint leading to multiple site damage, where many adjacent rivet holes may all have fatigue cracks of less than detectable size – perhaps 1–2 mm at the 90/95 level. Collectively these undetected cracks can act to reduce the ultimate strength of the structure when they are present ahead of the tip of a single long crack (Figs 15 and 16).

Criteria for propagation of long cracks in the presence of multiple site damage have been intensively investigated. Swift[9, 32] has proposed a simple criterion which provides an approximate solution. Swift proposes that the long crack tip and the next crack link when their plastic zones coalesce (Fig. 17) . More complex criteria have been proposed by Newman[33] involving critical crack tip opening angles. The simple Swift criterion does provide approximately correct answers compared with experiments, and it confirms that residual strength levels will be reduced in the presence of MSD.

Rivet holes with small cracks

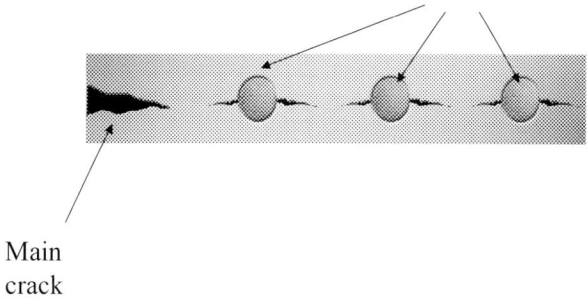

Main
crack

Fig. 15 Multiple site damage in rivet array.

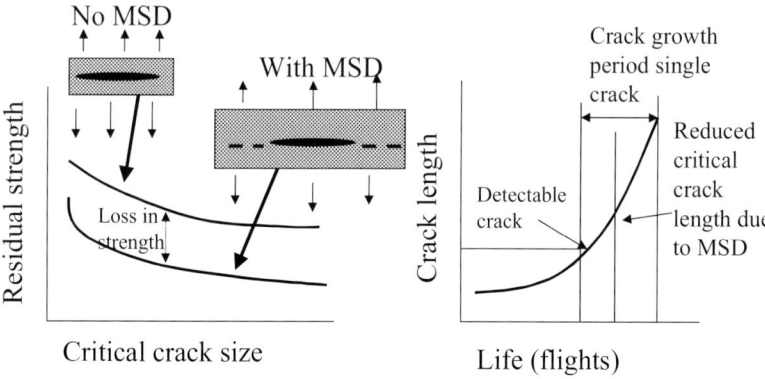

Fig. 16 Loss in strength caused by multiple site damage (adapted from Ref. 14).

In terms of the damage growth diagrams used earlier, the MSD situation is illustrated in Fig. 18 which shows residual strength in the upper half and damage growth below. The reduction in strength due to local single crack discrete damage growth is shown, together with the damage detection limits which provide the safety net against inadvertent failure. In the absence of discrete damage, the development of MSD is also shown, occurring at first much more gradually, but producing in the end a more rapid degradation in residual strength. The result is a greatly reduced period for which MSD is detectable prior to failure, i.e. when the residual strength of the structure is no longer able to support the design limit load.

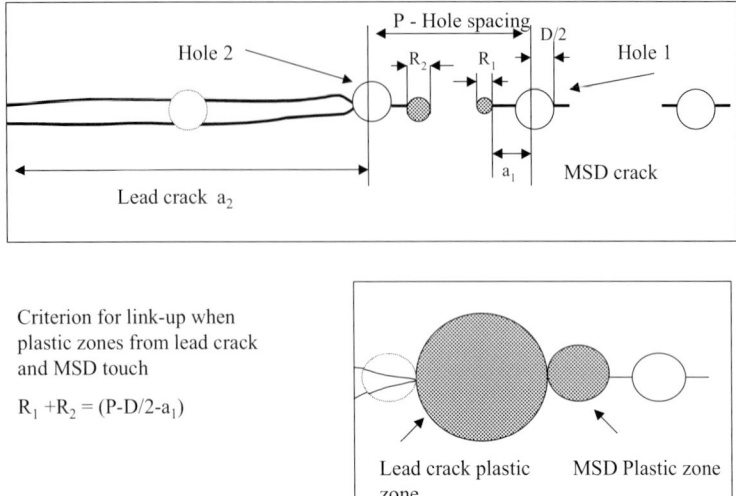

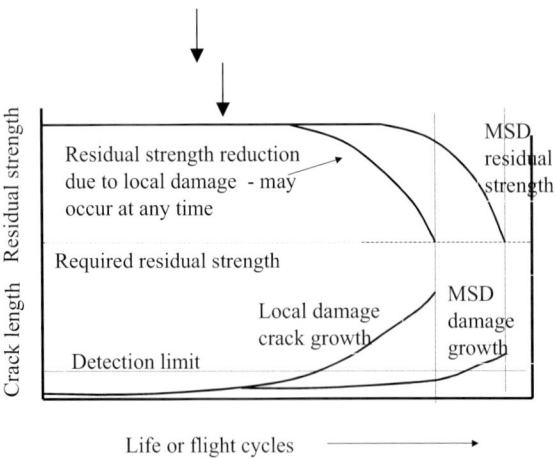

Fig. 17   The Swift approximate criterion for link up of MSD cracks with lead crack.[28]

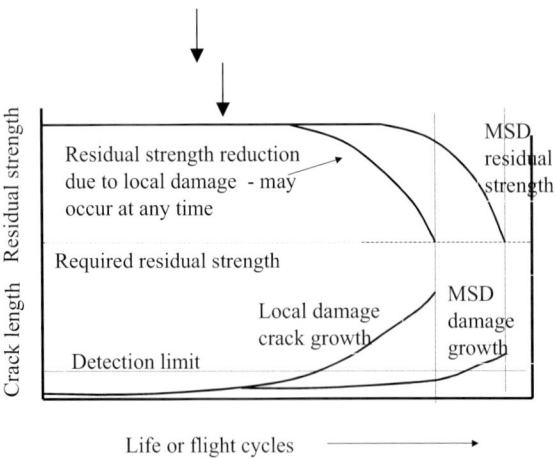

Fig. 18   The poor detection characteristics of MSD damage, combined with the reduction in residual strength means that when MSD has attained detectable size, there is little life remaining up to the point where residual strength cannot resist design limit load.

Because of the difficulties of detection of MSD in riveted joints, it is becoming clear that, in the future, regulators will not permit operation of aircraft in regions of life where development of widespread fatigue damage is likely. A recent proposal by the FAA[24] has suggested that aircraft certified after 1998 should be subject to this condition. The design service goal of an aircraft would be defined as the life during which MSD/WFD is not expected to occur. Operation beyond this point would be subject to stringent conditions for inspection and assessment of structural integrity. At the time of writing, this proposal is still in discussion between the European JAA and the US FAA.

COMPOSITE MATERIALS

Carbon fibre reinforced polymers (cfrp) composites have been used on secondary parts of the aircraft structure for many years. Their use is slowly spreading from secondary parts – e.g. engine cowlings, fairings; to primary and safety critical parts of the aircraft structure. For example, in addition to composites in many other parts of the aircraft, the Boeing 777 has the whole of the empennage (tail fin, tail plane and rear fuselage) in cfrp. Airbus aircraft also have tail fin and tail plane in cfrp. The coming A380 aircraft, the largest passenger aircraft yet built, is planned to have the wing carry through box in cfrp, as well as the tail plane and fin. Other parts of the wing primary structure will follow. The reason for the advance of composites is their excellent stiffness-to-weight and strength-to-weight properties. However their damage tolerance performance reduces these advantages considerably. Because of this, design strains in composite components are currently little different to those found in aluminium structures.

It is clear that cfrp is a very different material from either 2xxx or 7xxx aluminium alloys, and its mechanical behaviour, particularly aspects related to damage tolerance, are also very different. While the basic requirements for damage tolerant certification remain unchanged – composite aircraft structures for large transport aircraft are still covered by 25.571 – the interpretation, emphasis and tests required to demonstrate compliance with 25.571 are very different. Composite structures are covered by separate advisory material[34] on acceptable means of demonstrating compliance.

The aspects of cfrp mechanical behaviour which are significant from the damage-tolerance point of view include:

(1) The poor through-thickness strength and fatigue performance, in comparison with the in-plane performance
(2) The lack of a region of plasticity after yield, prior to final failure, which is common to all metals.

(3) The response to out-of-plane impact (from stones, birds, hail, dropped maintenance tools etc).
(4) The relatively poor properties in compression, in comparison with tension, (whereas metals have superior compression properties compared with those in tension).
(5) Environmental degradation of mechanical properties with temperature and with absorption of water from the surrounding environment.

There is the further difficulty that when composite structures are manufactured, the material is created at the same time as the structure is fabricated. For metals, the material is manufactured and then the structure or component is machined or fabricated from it. With metallic materials there will of course be some degradation from small samples to large components because of the well-documented phenomenon of size effects in fatigue and static strength. In this case, the small coupons may be cut from the large component; sampling the same material. In the case of the composite, it is not necessarily the case that the material of small coupon samples will be the same as that of the material in the final wing or fuselage; the properties may consequently be different for this reason, as well as for size effect reasons.

Thus a major concern in fatigue substantiation of composites, is demonstration that properties achieved at the coupon component or sub-component level will be the same as those operating in the complete structure. This necessitates an expensive pyramid of testing, (Fig. 19) designed to demonstrate the transferability of properties from coupon to structure.

Both in-plane static and fatigue properties of cfrp are excellent (typical static strength for quasi isotropic cfrp is 700–900 MPa) . However in laminates, because

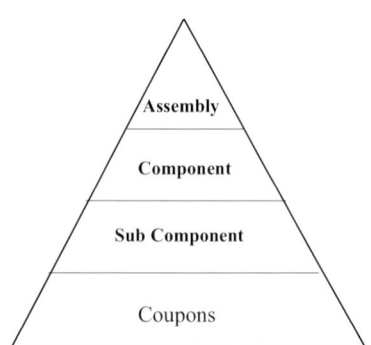

Fig. 19   Composite material test pyramid.

of the elastic mismatch between adjacent layers with different orientations, design features such as free edges, holes or changes in thickness all result in the development of through-thickness stresses as a consequence of in-plane loadings. The through-thickness strength of the laminate is limited by the resin, which has strength values at most of about 100 MPa. Thus small through thickness stresses may readily nucleate delamination cracks. In-plane delaminations, resulting from resin fracture between the carbon fibres, can propagate through the sample in response to static or fatigue loading in the out-of-plane direction. Similar damage can result from in-plane shear stresses, where delaminations are propagating in mode II shear.

Delaminations may also result from low velocity impacts of quite moderate severity. Impacts of 10–15 J may produce cones of delamination damage extending out and down from the impact point (see Fig. 20). The diameter of the damage will depend on the impact energy, the thickness of the laminate and the toughness of the resin.[35] However the delamination damage is caused, the influence on in-plane compression strength is much more severe than the influence on in-plane tension

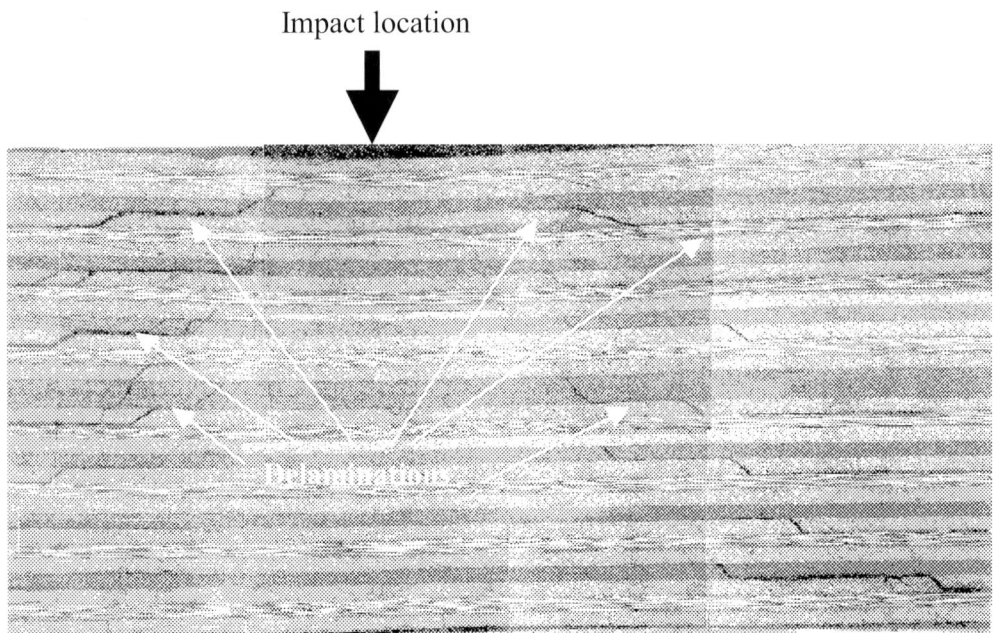

Fig. 20 Cross section through impact damaged area of multi-layer carbon fibre laminate Hexcel (T300/914) showing interlayer and cross layer delaminations and cracks; the impact site is at the indentation in the top surface ($\times 20$).

properties. In tension, if there is little fibre damage resulting from the impact; load carrying capacity of the fibres is scarcely affected by the presence of delaminations. In compression, compressive buckling of fibres around the delaminated area takes place, greatly reducing laminate in-plane compression strength.

To optimise compression performance of impact damaged laminates, the 'compression after impact' (CAI) test[36] has been developed. A small coupon of cfrp, 100×150 mm in size is impacted and the damaged coupon is then subjected to in-plane compression loading to failure. This test is used at coupon level to rank materials for their ability to resist both impact and compression. The test result is not a material property as such; results will depend on the impact energy and thickness of sample, as well as being a function of two separate properties (a) the impact damage resistance and (b) the compression strength in the presence of that impact damage. Typical values of compression after impact strength are in the region of $250–300$ MPa[37] – this imposes a design limitation of about 0.3–0.4% strain on the structure.

After impacts of low energy, there may be internal delaminations in the laminate interior, but no visible damage on the surface. The critical energy level at which damage becomes first visible on the surface is called the Barely Visible Impact Damage (BVID) level. From a damage tolerance point of view, it must be assumed that such damage exists all the time, and the relevant mechanical properties to be used in damage tolerant design are those of the composite in the presence of BVID damage levels. Hence static compression strength will be that achievable in a CAI test with BVID present. Fatigue strength will be the alternating stress required for propagation of delaminations from impact damage formed at BVID impact energy levels.

There has been extensive research into test techniques to measure and define conditions for delamination growth under static and fatigue loading.[38, 39] There now exist standard tests to determine G_{IC} and G_{IIC} critical strain energy release rates for delamination growth under static loading in modes I and II. Corresponding tests under fatigue loading have demonstrated that delamination fatigue crack growth may be correlated with ΔG, the strain energy release rate range, in an analogous test to the fatigue crack growth rate test to characterise metallic fatigue crack growth rates with ΔK the stress intensity range.

In metallic materials, there is a wide range of stress intensity values over which crack growth can take place, and moderate exponents of 2–4 in the relation between stress intensity ΔK and growth rate da/dN. This leads to a wide design window in which for many design situations, damage tolerance can easily be demonstrated. In CFRP materials this window is small. Exponents for delamination crack growth are double or triple those found in metals.[39] Delamination crack growth rates are fast and acceleration rapid, leading to a small design window in which it is very difficult to demonstrate damage tolerance.

For example[39] the laminate 914/T300 has an exponent of 14–16 in mode I delamination fatigue crack growth, using the expression $\Delta G = \Delta K^2/E$ to establish equivalences between G and K in order to facilitate comparison of composites with metals. A comparison of fatigue crack growth rates for mode I delamination crack growth in 914/T300 and aluminium 2024 T3 can be seen in Fig. 21.

In the delamination crack growth samples, idealised single delaminations with simple linear crack fronts are tested. Under these circumstances both in shear mode II and opening mode I, thresholds for fatigue delamination crack growth are low e.g. ΔK values[39] less than 1 MPa m$^{1/2}$. In real structures, impact and other forms of delamination damage have complex crack fronts and have many different misoriented delaminations. The effect of this is to raise the threshold for fatigue delamination growth from impact damage to a significant fraction (0.7–0.8) of the static in-plane pristine compression strength. In this sense composite materials have good damage tolerance. However, once growth begins, its rate is rapid, and the number of cycles from initiation of growth to final failure remain small. Damage tolerance via a sub-critical crack growth methodology cannot practically be demonstrated.

There is little published data relevant to this real damage growth situation. Data are required on samples of cfrp, impact damaged to BVID levels and tested in in-plane compression fatigue under constant and variable amplitude loading. Some recent published results[40] which confirm the general picture described above, are summarised in Fig. 22.

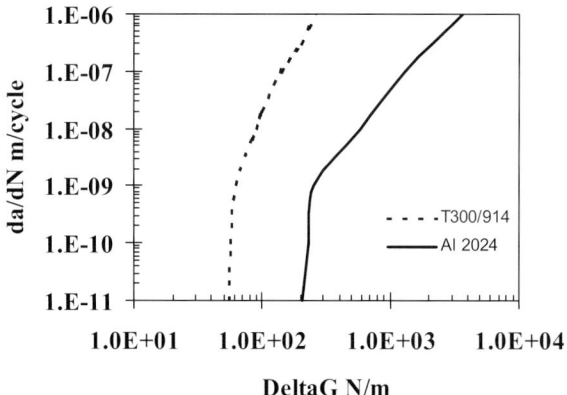

Fig. 21 Comparison in terms of ΔG of crack growth rates in unidirectional T300/ 914 epoxy-carbon fibre composite, with 2024 aluminium alloy, using the expression $\Delta G = \Delta K^2/E$ to convert. Data at $R = 0.1$ for both test series.

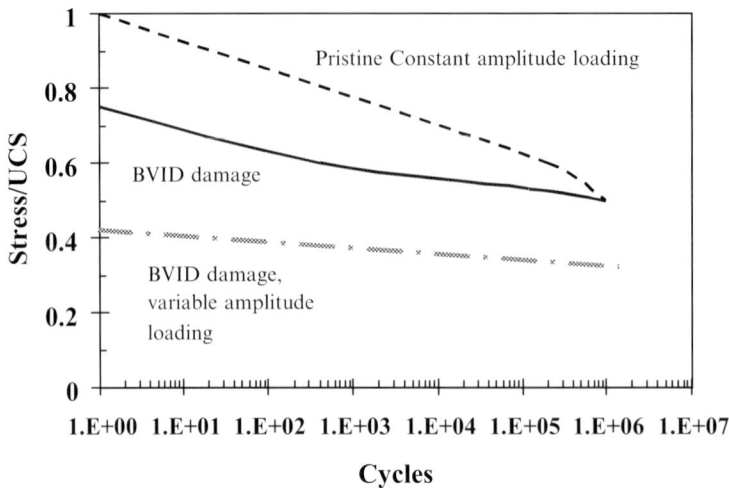

Fig. 22 Fatigue performance of damaged and undamaged laminates in compression, from Mitrovic.[36]

Figure 22 shows that the static compression strength in a quasi-isotropic $\{0°; \pm45°; 90°\}$ laminate in cfrp containing impact damage at BVID levels, is reduced to about 75% of the pristine laminate strength. Constant amplitude fatigue cycling in compression further reduces this to about 60% of the original compression strength. Below 60% of the pristine compression strength there is no damage growth in either damaged or undamaged laminates. Variable amplitude cycling, with compression overloads reduces the fatigue compression strength still further. As these lines have such a shallow gradient, the transition between zero damage growth and a failed structure will be rapid and will occupy relatively few fatigue cycles.

This behaviour may be contrasted with that of metallic materials (Fig 10(a)). Here the worst case is of course in tension, and the presence of an assumed level of initial damage (generally a 1.25 mm flaw), will have little effect on the static strength, but after prolonged fatigue cycling, damage growth can occur at a stress as low as 10% of the static strength for aluminium alloys, and an even smaller fraction of the static strength of titanium or high strength steels (Fig. 10(b)). In the undamaged state, high cycle fatigue strengths in aluminium alloys are about 0.3 of the static strength.

This difference in degradation behaviour leads to a very different plot of residual strength against service life for cfrp laminates (Fig. 23) Whereas metallic structures will have a gradual degradation with life, the compressive strength of the composite one will remain unchanged until such time as impact damage

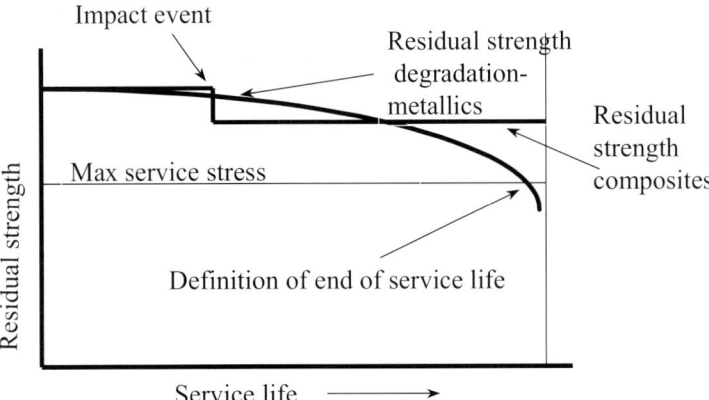

Fig. 23 Degradation in residual strength, composite and metallic aircraft structures.

occurs – at which time the residual strength will reduce abruptly, and then remain unchanged until further impacts occur. Instead of a damage growth diagram with a wide design window, and a gradual increase in damage level, there will be a series of irregular and unpredictable increases in damage, not directly related to the time the structure has been in service. Effective prognosis of the future damage level and lifetime of the structure is not readily possible, as the service period required for a transition from a safe state to a failed state cannot be predicted as it can for metallic structures.

To demonstrate damage tolerance, CFRP structures hence have to rely on demonstrations of zero damage growth from impact damage at BVID levels throughout the lifetime of the structure – a general guideline is no damage growth in twice the design service goal. For composite structures there is increased emphasis on degrading the structure in terms of fatigue impact damage and environmental degradation prior to static testing. This is because of the increased influence which small damage levels have on static strength, particularly in compression.

There are obvious difficulties in defining an inspection regime (how often, with what sensitivity?) in such a situation. Damage growth calculations based on the results of delamination crack growth tests would result in very short, uneconomic inspection intervals. More realistic approaches would be based on the likelihood of detectable impact damage having occurred. Clearly a continuous monitoring system for damage would be the best option, and there are intensive efforts proceeding world wide to develop an effective health monitoring system for damage detection in composite aircraft structures.

FUTURE PROCESS AND INSPECTION DEVELOPMENTS

In the move towards reduced cost of manufacture of large transport aircraft, a number of cost saving process options are being considered. These include use of casting and large forgings instead of fabricated structures. An example of the use of castings is the development of a cast aircraft fuselage door,[41] replacing a complex riveted fabrication. The use of welding instead of riveting for parts of the aircraft fuselage and wings is also under active development.[42] All of these processes create large integral structures, with an easy crack path from one part of the structure to another. Mechanically fastened structures are considered[43] to have better damage tolerance than integral structures because of the difficulty of propagating cracks from one component to another. Considerable research will be required to ensure that safety is not compromised by these new manufacturing developments.

Application of the damage tolerance approach will continue to underpin design against fatigue and other degradation processes in aircraft service. Future developments may well see an extension in the application of service damage detection and HUMS to other components of the helicopter, and into fixed wing aircraft, particularly in support of the use of composite materials, where rapid detection of service impact damage is essential.

ACKNOWLEDGEMENTS

Thanks are due to many colleagues for helpful advice and discussions. Particular thanks are due to J. Bristow, Head of the Structures & Materials Department at the UK Civil Aviation Authority. His advice, encouragement and support has been invaluable.

REFERENCES

1. 'World Aircraft Accident Survey', *Flight International*, January 2000.
2. *Global Fatal Accident Review*, 1980–1996, CAP 681, Civil Aviation Authority, 1998.
3. 'Airplane disaster caused by Fatigue of Steel', *Iron Age*, Sept. 1929, **124** (10), p. 600.
4. A. Palmgren, 'Die Lebensdauer von Kugellagern', *Zeitschrift des Vereins Deutscher Ingenieure*, 1924, **68**, pp. 339–341.
5. R. B. Bland and P. E. Sandorff, 'The control of life expectancy in airplane structures', *Aeronautical Engineering Review*, 1943, **2** (8), pp. 7–21.
6. M. A. Miner, 'Cumulative damage in fatigue', *Journal of Applied Mechanics*, 1945, **12**, pp. 159–164.

7. *CAP 127 Civil aircraft accident report into the accidents to Comet G-ALYP and Comet GALYY*, HSMO, 1955.
8. J. Bristow, 'The Meaning of Life', in *Proc. CEAS Forum on Life Extension – Aerospace Technology Opportunities*, Cambridge, Royal Aeronautical Society, 1999, pp. 24.1–24.8.
9. T. Swift, 'Damage Tolerance in Pressurised Fuselages', *Proc. 14th ICAF, 1987: New materials and fatigue resistant aircraft design*, EMAS 1988, pp. 1–77.
10. British Civil Airworthiness Requirements, Section D, Issue 3, July 1956, Chapter D3-7.
11. P. C. Paris, M. P. Gomez and W. P. Anderson, 'A rational analytic theory of fatigue', *The Trend in Engineering*, 1961, **13**, pp. 9–14.
12. Joint Aviation Requirements JAR 25.571, Fatigue requirements for Large aircraft, first issue 1974.
13. J. P. Gallagher, F. J. Geissler, A. P. Berens, R. M. Engle and H. A. Wood, *USAF Damage Tolerant Design Handbook*, AFWAL-TR-82-3073, US Dept Commerce NTIS, 1984.
14. W. D. Rummel, P. H. Todd, S. A. Freska and R. A. Rathke, *The Detection of Fatigue Cracks by Non Destructive Methods*, NASA CR 2369, NASA, 1974.
15. U. G. Goranson, C. K. Gunther and W. T. Hardrath, 'Residual strength characterisation of jet transport structures', *Proc. 13th ICAF, Pisa, 1985: Durability and Damage Tolerance in Aircraft Design*, EMAS, 1985 pp. 585–604.
16. T. Swift, 'Effect of multiple site damage on certified lead crack residual strength', in *Proc. CEAS Forum on Life Extension: Aerospace Technology Opportunities*, Royal Aeronautical Society, 1999, pp. K1–K33.
17. ESDU data items 81031, 82015, 84003, Engineering Sciences Data Unit, ESDU International plc, London, 1983.
18. AFGROW Version 4.0001.11.8 18 May 2000, available at www://fibec.flight.wpafb.af.mil/fibec/afgrow.html.
19. C. Boller and T. Seeger, *Materials Data for Cyclic Loading*, Elsevier, 1987.
20. Joint Airworthiness Regulations, section 29, Requirements for large helicopters.
21. J. W. Lincoln, 'Damage tolerance for helicopters', *Proc. 15th ICAF, Jerusalem*, EMAS, pp. 263–290.
22. 'Rotorcraft Transmission System Health monitoring – Experience update'; Aerospace Industries Division Seminar, co-sponsored by RAeS; November 1997, I Mech E, 1997.
23. J. McColl, 'Overview of transmissions HUM Performance in UK North Sea Helicopter Operations', paper 2 in *Rotorcraft Transmission System Health Monitoring – Experience update*, Aerospace Industries Division Seminar, co-sponsored by RAeS, November 1997, I Mech E, 1997.
24. P. E. Irving and R. A. Hudson, 'Routes to damage tolerance and life extension in helicopters and other high strength high integrity mechanisms', *Proc. Fatigue 2000: Fatigue durability of materials components and structures*. M. R. Bache, P. A. Blackmore, J. Draper, J. H. Edwards, P. Roberts and J. R. Yates eds, Cambridge, April 2000, EMAS, 2000.
25. 'Survey of Ageing Aircraft', *Flight International*, June 2000.

26. J. W. Bristow, 'The Age of the Plane', Presentation to Institute of Mechanical Engineers, London, April 1993.
27. Federal Aviation Regulations, FAR 25 Para 571, amendment 96, March 1998.
28. Department of Trade Aircraft accident report 11/77, HMSO, 1977.
29. Department of Trade Aircraft accident report 9/78, HMSO, 1978.
30. *FAA/NASA International Symposium on Advanced Structural Integrity Methods for Airframe Durability and Damage Tolerance*, NASA Conference Publication 3274, NASA, 1994.
31. R. M. Burynski, G. Chen and R. P. Wei, 'Evolution of pitting corrosion in a 2024 T3 Aluminium alloy', ASME international Mechanical engineering congress and exposition 1995, San Francisco, *Proc. Structural Integrity in Aging Aircraft*, AD-vol. 47, ASME, 1995.
32. T. Swift, 'Widespread fatigue damage monitoring – Issues and concerns', *FAA/NASA International Symposium on Advanced Structural Integrity Methods for Airframe Durability and Damage Tolerance*, NASA Conference Publication 3274, NASA, 1994, pp. 829–851.
33. B. R. Seshadri, J. C. Newman, D. S. Dawicke and R. D. Young, 'Fracture analysis of the FAA/NASA wide stiffened panels', *Proc. 2nd Joint NASA/FAA/DoD Aging Aircraft Conference*, Williamsburg VA, NASA, 1998.
34. Joint Airworthiness Authorities ACJ 25.603; Acceptable means of compliance for composite aircraft structure.
35. F. L. Matthews and R. D. Rawlings, *Composite Materials: Engineering & Science*, Chapman & Hall, 1994, pp. 363–373.
36. Boeing Specification support standard, 'Advanced composite compression test', BSS 7260. Boeing Airplane Company, 1986.
37. R. J. Cano and M. B. Dow, 'Properties of five toughened matrix composite materials', NASA T-P 3254, NASA, 1992.
38. F. Ozdil and L. A. Carlsson, 'Mode I Interlaminar Fracture of Interleaved Graphite Epoxy', *J. Composite Materials*, 1992, **26**, pp. 432–459.
39. M. Hojo, K. Tanaka, C. G. Gustafson and R. Hayashi, 'Effect of stress ratio on near threshold propagation of delamination fatigue cracks in unidirectional CFRP', *Composites Science & Technology*, 1987, **29**, pp. 273–292.
40. M. Mitrovic, H. T. Hahn, G. P. Carman and P. Shyprykevich, 'Effect of loading parameters on fatigue behaviour of impact damaged composite laminates', *Composite Science & Technology*, 1999, **59**, pp. 2059–2078.
41. S. F. Kenner Knecht, G. Witgens, R. Tombori, P. van Biljn and X. Durant, 'Commercial development of D357 alloy investment cast aircraft door substructures', *Proc. International Non-Ferrous Processing and Technology Conference*, St Louis, March 1997, ASM International, 1997.
42. P. E. Irving and G. Bussu, 'Damage tolerance of welded aircraft structures', *Proc. 21st ICAF*, Toulouse, 2001.
43. T. Swift, 'Application of damage tolerance technology to type certification', SAE technical paper 811062, October 1981.

CHAPTER 6

Recent Developments in Methodology for Fracture Mechanics Assessments

D. P. G. Lidbury

Structural Integrity Department, Engineering Integrity Group, Serco Assurance, Risley, Warrington, WA3 6AT, UK

ABSTRACT

Fracture mechanics is a discipline that seeks to predict the behaviour of bodies containing crack-like (planar) defects under load. Engineers routinely carry out defect assessments to assess the integrity or fitness for purpose of structures or components whose failure would have unacceptable safety or economic consequences. The development and validation of defect assessment methodology is therefore an important aspect of fracture mechanics research. The predictions of a given methodology are validated by comparing them with failure data from large-scale tests specially designed to simulate specific structural features. Within this context, two recent international programmes have helped to validate defect assessment methodology:

- NESC 1 Spinning Cylinder Test (simulation of reactor pressure vessel integrity under pressurised thermal shock conditions)
- BIMET (structural integrity of dissimilar metal welds)

Each of these has considered specific structural features to investigate the transferability of data obtained from tests on laboratory-scale specimens in predicting the fracture behaviour of components under complex conditions.

The purpose of the present paper is to describe these programmes and their main results, and highlight their benefits. In particular, it is shown that although the programmes address issues that are important to the nuclear power industry, their results are relevant to the development of defect assessment methodology more generally.

INTRODUCTION

The use of fracture mechanics to determine the integrity or fitness for purpose of structures or components containing planar defects (cracks) is an important aspect of safety assessment. In any engineering fracture mechanics assessment, the stability of a postulated or known crack is assessed by comparing load and resistance

103

terms. Proximity to crack initiation is determined by calculating the crack driving force as a function of applied load, crack shape and geometry and comparing it with the material fracture toughness, for the relevant temperature and loading rate.

Experimental validation is an important aspect of gaining acceptance for a defect assessment methodology. Here, the predictions of a given methodology are compared with failure data from large-scale tests specially designed to represent specific structural features. These tests are designed to simulate as closely as possible the behaviour of cracks in components under conditions representative of normal or abnormal loading.

In the nuclear industry, two areas where it has proved valuable to carry out structural features tests to help validate safety assessments relate to (i) crack behaviour under design basis accident conditions involving pressurised thermal shock (PTS); (ii) the defect tolerance of dissimilar metal welds (DMWs). With respect to (i), PTS scenarios result in the reactor pressure vessel (RPV) of a light water reactor (LWR) experiencing complex conditions of thermal shock and pressure loading. These loadings are multiaxial and vary with time into the transient. Shallow surface-breaking/near-surface cracks are most likely to propagate under PTS conditions. This is because the thermal stresses generated by activation of the reactor emergency core cooling system in response to some initiating event (e.g. a steam line break) are highest close to the inner surface of the RPV. Also, the RPV inside wall is the surface closest to the reactor core, and therefore suffers the greatest degradation of toughness due to in-service exposure to neutron irradiation. With respect to (ii), DMWs are used in LWRs to join heavy-section ferritic steel components to austenitic stainless steel pipework. In the case of pressurised water reactors (PWRs), DMWs are used to join primary circuit piping to the RPV nozzles. Here, concerns about the fracture behaviour of BMWs arise from service experience of cracking at the outer surface in the vicinity of the ferritic-austenitic interface – e.g. see Ref. 1. This is a region where material characteristics are not easily defined due to the rapid transition in metallurgical structure. Moreover, mechanical analyses are made more complex by the presence of residual stresses and mixed-mode loading. Defect tolerance is an important safety issue, since failure of a DMW by propagation of a crack from the outer surface to cause leakage or rupture would constitute a breach of the primary circuit pressure boundary.

Within the above context, the following international programmes have addressed various aspects of defect assessment methodology:

- NESC-1 (simulation of reactor pressure vessel integrity under pressurised thermal shock conditions)
- BIMET (structural integrity of dissimilar metal welds)

Each of these programmes has considered the simulation of specific structural features in order to investigate the transferability of data obtained from tests on laboratory-scale specimens in predicting the fracture behaviour of components.

The following sections describe the above programmes in turn from the point of view of developments in methodology for fracture mechanics assessment. A concluding discussion then assesses their benefits for the development of defect assessment methodology, both for the nuclear power industry and more generally.

NESC-1 SPINNING CYLINDER PROJECT

The Network for Evaluating Steel Components (NESC) was formed to bring together experts, primarily within Europe, to examine and develop methods for assessing the structural integrity of nuclear power plant (NPP) components. Operated by the European Commission's Joint Research Centre at Petten, The Netherlands, NESC is concerned with examining all aspects of structural integrity assessment and their integration.

The NESC Spinning Cylinder Project was launched in 1993 as the first major project of the Network, with the test itself sponsored by the UK Health and Safety Executive (HSE). The aim of the project was to address the technical issues arising from the entire process of structural integrity assessment – inspection, prediction and testing.[2] To this end its principal objectives were to:

- Assess different inspection methods for detecting and sizing through-clad and underclad defects in a thick-walled steel cylinder before and after it had been subjected to a simulated pressurised thermal shock transient
- Determine experimentally the behaviour of underclad and through-clad defects in a thick-walled steel cylinder subjected to a simulated pressurised thermal shock transient
- Benchmark different fracture mechanics methods used to predict the behaviour of underclad and through-clad defects under pressurised thermal shock conditions.
- Gain a fuller understanding of the interaction between the disciplines of materials evaluation, non-destructive testing and fracture analysis

These objectives have been addressed by NESC through the creation of Task Groups, which respectively cover the areas of Inspection, Materials, Structural Analysis, Instrumentation and Evaluation.[3]

The spinning cylinder test facility (Figure 1) was designed to investigate the behaviour of cracks in heavy-section steels under loading conditions representative of postulated pressurised thermal shock (PTS) transients in pressurised water

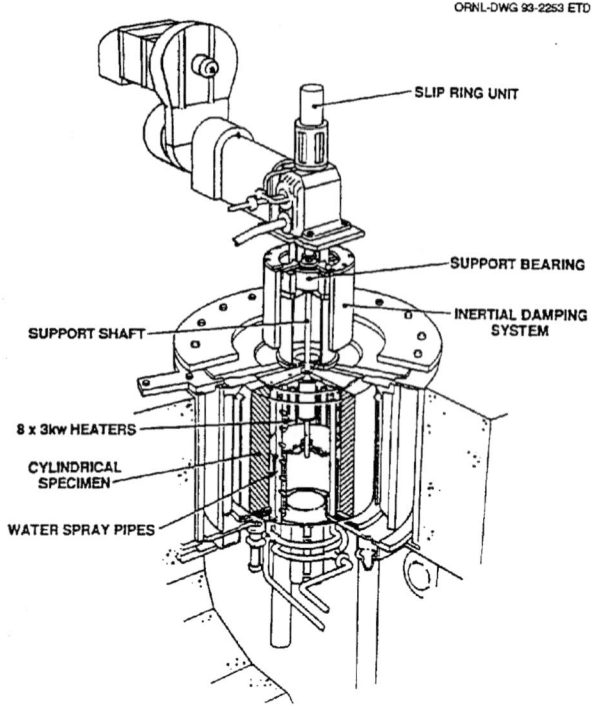

ORNL-DWG 93-2253 ETD

Fig. 1   Schematic details of spinning cylinder test facility.[3]

reactor (PWR) reactor pressure vessels (RPVs). Various combinations of centrifugal and thermal shock loading are used to simulate a PTS transient. Primary loading is applied to the hollow test cylinder by rotating it at high speed (up to 2750 rpm) about its longitudinal axis. Thermal shock loading is applied independently by spray quenching the defect containing inner surface of the cylinder (which is pre-heated typically to 290°–300°C) with water at ambient temperature. Prior to the NESC Spinning Cylinder Project, AEA Technology had carried out six tests on unclad cylinders at its Risley site (1). Two of these tests (No. 4 and No. 6) investigated the cleavage and arrest behaviour of fatigue sharpened surface-breaking defects inserted into unclad cylinders.

NESC-1 was set up to provide data on the behaviour of various underclad defects and a large, through-clad defect, under simulated PTS loading [4]. The test piece consisted of an internally clad, 7-tonne A508 ferritic steel cylinder, of outer diameter 1,395 mm and total wall thickness 175 mm. The type 316 austenitic stainless steel two-layer cladding was machined to a final thickness of 4 mm. The cylinder material was specially heat treated to be representative of an ageing PWR RPV.

Sixteen defects, differing widely in fabrication method, size and location, were introduced into the cylinder. The largest of the underclad defects (Defect B) was initially 261 mm long and 77 mm deep; the dimensions of the corresponding large through-clad defect (Defect R) were 208 and 78 mm respectively (see Fig. 2). Further details of the various defects may be found in the publication by Bass, et al . [4] The spinning cylinder test was performed in March 1997. The observed crack growth behaviour of the defects B and R was as follows:[4,5]

DEFECT B

Crack extension occurred along the entire front of Defect B. A maximum crack extension of 15 mm was observed just below the cladding heat affected zone (HAZ). At the deepest point of the defect, the crack extension was up to 4.5 mm. At one end of the defect the mechanism of extension was primarily intergranular cracking, whereas at the opposite end it was initially intergranular, followed by ductile tearing. The ductile tearing mode became more predominant

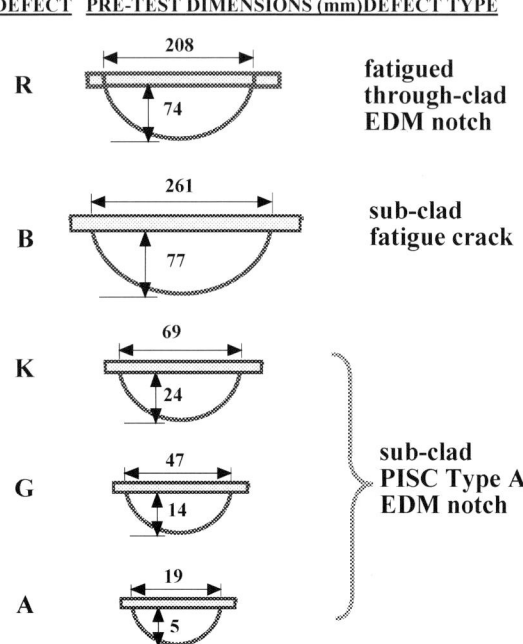

DEFECT	PRE-TEST DIMENSIONS (mm)	DEFECT TYPE
R	208 / 74	fatigued through-clad EDM notch
B	261 / 77	sub-clad fatigue crack
K	69 / 24	sub-clad PISC Type A EDM notch
G	47 / 14	
A	19 / 5	

Fig. 2 Defect identification and dimensions used in pre-test analyses of NESC-1 spinning cylinder test.[3]

at deeper positions along the crack front. The observation of a large stretch zone indicated that a considerable crack driving force was required to initiate crack extension.

DEFECT R

Crack extension occurred along the entire front of Defect R. At one end of the defect, the initial extension was followed by a local cleavage run-arrest event that occurred in the region just below the clad HAZ at between 213 s and 217 s into the thermal transient. In that region, the maximum crack extension was approximately 17 mm. The initiation site for this cleavage event was approximately 16.5 mm below the clad surface. Post-test examination of trepanned sections of the cylinder by scanning electron microscope (SEM) confirmed that the pre-cleavage crack extension was by ductile tearing. In the region of maximum crack depth, the crack extension was irregular, with small pockets of cleavage fracture approximately 1 mm deep.

SUMMARY OF PRE- AND POST-TEST FRACTURE MECHANICS ASSESSMENTS

A significant effort was made in performing both pre- and post-test fracture analyses. The majority of this was focussed on the fracture behaviour of defects B and R. Overall, the techniques employed varied from code-based procedures and simplified engineering assessment methods (EAMs), to 3-D finite element models featuring non-linear constitutive equations and finite strain theory.

Primarily as a result of the thermal shock loading applied to the cylinder, those parts of the large cracks with the highest likelihood of crack extension were in the near surface regions. This introduced a number of structural features that complicated the fracture assessment:

- Cladding effects, including cladding residual stresses
- Gradients in tensile and fracture toughness properties
- Crack-tip constraint effects
- Effective crack front length
- Warm prestressing (WPS)
- Crack arrest

Overall, some 34 organisations contributed to the various pre- and post-test analyses of NESC-1. The main results and conclusions were as follows:

1. All of the pre-test analyses predicted that Defect R would extend in an axial direction, following initiation in base metal immediately beneath the

cladding HAZ. The analyses predicted various amounts of stable crack growth prior to cleavage. Similarly, predicted times for the cleavage event varied. The EAMs produced relatively conservative results; the more detailed 3-D finite element models allowed accurate predictions of both initiation time as well as the extent and location of cleavage fracture. It is noteworthy that none of the analyses predicted crack growth by cleavage in the radial direction.

2. The fact that predictions of a cleavage event in relation to Defect B did not occur was to some extent mitigated by the unforeseen intervention of crack growth by intergranular fracture.

3. In terms of the overall consistency and accuracy of the various analytical predictions, the major source of uncertainty was considered to be the intrinsic variability in fracture toughness (as assessed by small-specimen tests). Variations in the estimates of crack driving force were considered to have had a smaller effect.

4. As part of the post-test assessment, six of the code-based procedures in force within Europe and the USA were used to assess the maximum size of defect that could be allowed to remain in the cylinder. This was on the basis that the cylinder was considered as if it were a safety-significant component in a nuclear power plant subject to periodic in-service inspection (ISI). These analyses indicated that only very small defects with through-wall depths in the approximate range 1 to 9 mm would be allowable. Without allowance for the anticipated effects of WPS, this range was reduced to 1 to 3 mm.

5. Possible reasons for the extreme conservatism in the code-based assessments were considered in detail. These were attributed to a combination of factors, notably the tendency for code-based procedures to: (i) overestimate values of crack driving force, (ii) underestimate effective toughness values, and (iii) apply substantial margins of safety to an already conservative analysis. With respect to (i) cracked-body analyses showed that in the near-surface region an effect of cladding is to reduce the crack driving force. Cladding is ignored in code-based assessments, leading to conservative assessments of the crack driving force at such locations. Another source of conservatism is the use of elastic analysis to estimate the contribution of (steeply declining) thermal shock stresses to crack driving force. With respect to (ii), the generally enhanced fracture properties of the cladding HAZ are ignored in code-based assessments, as is the effective increase in toughness due to the loss of crack-tip constraint with increasing loading near to surface-breaking locations.

6. Because no discernible crack growth occurred from the other defects introduced into the cylinder (which were all appreciably smaller than defects B and R), the NESC-1 project did not provide precise guidance on

the conservatism inherent in current code-based assessments of shallow, near-surface defects subject to PTS loading. (This being the situation of greatest practical concern.) Nevertheless, the test demonstrated that under simulated PTS loading conditions, underclad defects up to 77 mm in depth would not propagate in a cleavage mode. The beneficial effect of cladding in inhibiting cleavage initiation in near-surface crack locations was evident by comparison of the behaviour of Defect B with that of Defect R. This effect of cladding was also evident by comparison of the results of the NESC-1 test with previous tests on unclad cylinders, where cleavage of large pre-existing cracks occurred (spinning cylinder tests Nos. 4 and 6).

BIMET

BIMET (Structural Integrity of Bi-Metallic Components) was a three-year collaborative research programme carried out for Directorate General Research Technology and Development (DG-RTD) of the European Commission under the Euratom Fourth Framework Programme 1994–1998: 'Nuclear Fission Safety'.[6] The project involved the collaboration of eight European partners, with Electricté de France (EDF) acting as project co-ordinator.

The overall objective of the project was to develop methods of analysis describing the behaviour of defects in ferritic to austenitic DMWs. The particular geometry under study was a welded pipe assembly of mean diameter 143 mm, containing a bi-metallic girth weld joining A508 ferritic steel and type 304 stainless steel sections. A circumferential surface-breaking defect was inserted into the pipe in the 309L austenitic buttering layer immediately adjacent to the ferritic base material. Type 308L austenitic steel was used for the remaining layers of buttering and as the weld filler material. The pipe was subjected to four-point bending at ambient temperature. Two large-scale experiments were conducted with this configuration, each with the objective of producing significant stable ductile tearing. Figure 3 shows schematic details of the test assembly. Figure 4 shows details of the defect geometry used in the tests BIMET-1 and BIMET-2 respectively.[7]

In both tests stable ductile tearing took place in the austenitic steel buttering following crack initiation. Crack extension occurred in a plane perpendicular to the pipe surface toward the inclined ferritic-steel/austenitic-buttering interface. In each case maximum crack growth was achieved over the central region of the crack front, with $\Delta a_{max} = 2.4$ mm in the case of BIMET-1 and 8.6 mm in the case of BIMET-2. Crack initiation was assessed by the electrical potential drop method. For BIMET-1, the initiation moment M_i was judged to be in the range $150 < M_i < 165$ kNm. The most likely value of M_i was considered to be 154 kNm,

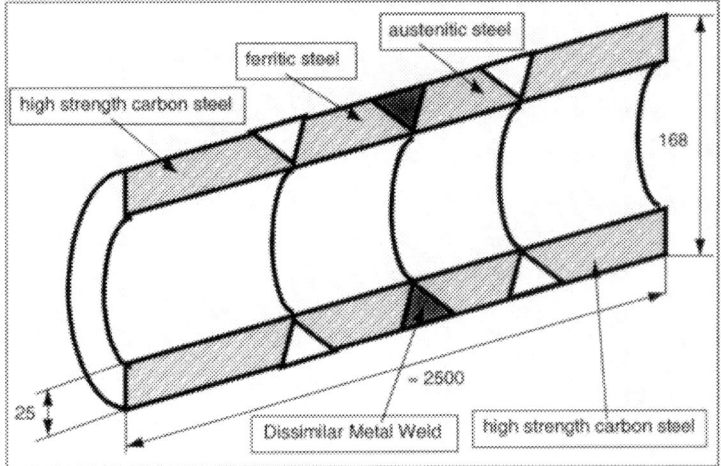

Fig. 3 Schematic details of DIMET test assembly.[6]

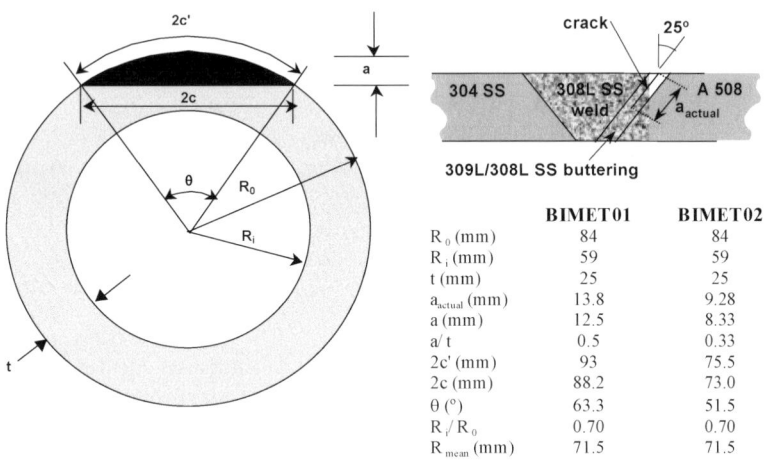

	BIMET01	BIMET02
R_0 (mm)	84	84
R_i (mm)	59	59
t (mm)	25	25
a_{actual} (mm)	13.8	9.28
a (mm)	12.5	8.33
a/ t	0.5	0.33
2c' (mm)	93	75.5
2c (mm)	88.2	73.0
θ (°)	63.3	51.5
R_i/R_0	0.70	0.70
R_{mean} (mm)	71.5	71.5

Fig. 4 Details of DIMET-1 and DIMET-2 defect geometry.[7]

and this was adopted as a reference value in judging the accuracy of the various test predictions[8] see Fig. 5. In the case of BIMET-2, the initiation moment was assessed as being somewhat higher and in the range $155 < M_i < 170$ kNm.

Apart from the analyses carried out by EDF to design the tests, all other analytical work was conducted post-test, with the majority of these analyses focussed on the behaviour of the test BIMET-1. Figure 5 summarises the results of the various analyses used in predicting values of M_i for BIMET-1. In all, seven

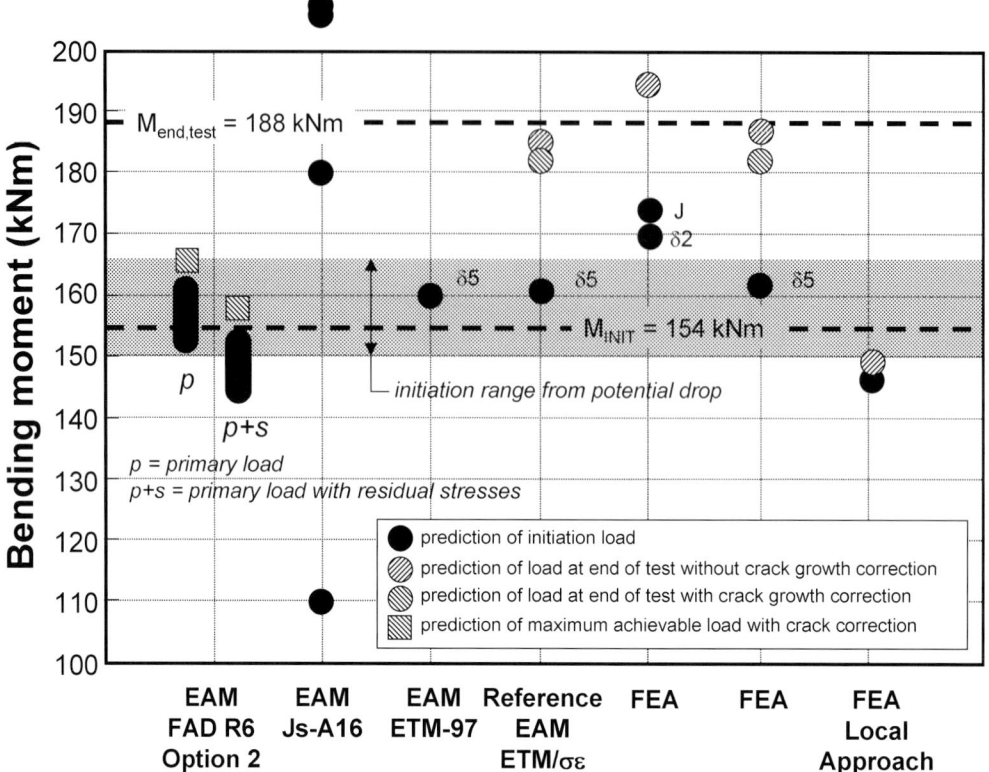

Fig. 5 Predicted values of applied bending moment at crack initiation for test DIMET-1.[8]

separate analyses were performed. Four of the analyses involved the use of simplified methods (EAMs). The remainder involved finite element analyses, with one of these associated with use of the Beremin local approach model of ductile fracture initiation (based on the well known Rice and Tracy void growth model), which was implemented as a post-processing step. While it is beyond the scope of the present paper to describe these analyses in detail, some general points can be made.

ENGINEERING ASSESSMENT METHODS

The various analyses based on EAMs demonstrate that the simplified fracture assessment procedures used can be successfully applied to a complex structural situation. This is subject to the structural features under consideration being suitably reduced to an idealised problem. Here, the following considerations are important:

- Describing the overall system and its boundary conditions appropriately
- Describing the crack geometry and its orientation with respect to the applied loads appropriately
- Selecting relevant handbook solutions for calculating stress intensity factors
- Selecting relevant handbook solutions for calculating limit loads and, if necessary, modifying these solutions in the light of relevant finite element analysis results
- Ensuring that gradients in material properties have been suitably assessed by correct interpretation of available data obtained from testing small specimens
- Recognising that failure assessment curves and weld mismatch correction factors can be very sensitive to the shape of the true stress vs. true strain curve, particularly close to the yield point
- Taking due account of the presence of residual stress

Figure 5 shows that three out of the four analyses based on EAMs produced very good estimates of the applied bending moment for crack initiation in the test BIMET-1. The results of the other EAM analysis are less satisfactory, but this is in part thought to be due to the particular selection and interpretation of materials data for that analysis. All analyses showed estimation of the plastic limit load (M_L) to be the most significant parameter in failure assessment, since for this problem crack initiation occurred at a load that was close to M_L.

FINITE ELEMENT ANALYSES

The three finite element analyses underlined the importance of using precise tensile and flow properties for the various regions of the test assembly. Here, appropriate selection of the curve fits to represent true stress vs. true strain data, including tensile properties of the 304-L austenitic steel, proved especially important in obtaining good agreement between the predictions and test results shown in Fig. 5. Crack initiation was determined by comparing crack driving force with either J- or CTOD-based resistance curves. The CTOD-based δ_5 approach had some advantages over the J-based approaches used in commercial finite element codes, the latter exhibiting path dependency due to significant crack-tip blunting and having the accuracy of correction terms for 'far-field' values restricted by the effect of material inhomogeneity. The resistance curve data were used to estimate values of $\Delta a_{\max}$. These values were all less than the observed value for BIMET-1 of 2.4 mm. (Values of Δa_{max} were also underestimated in the one EAM analysis in which predictions were made of the extent and stability of crack growth.) Lastly, two of the finite element analyses considered the effect of residual stress (measured experimentally as part of the BIMET project) on crack initiation under fully plastic conditions, and found this to be negligible. In contrast, one of the EAM analyses accounted for the presence of weld residual stress and found this to reduce estimated values of M_i by between 5 and 15%, the exact value depending upon other analytical assumptions – primarily the method used to estimate the plastic limit load.

CONCLUDING DISCUSSION

It is evident that there are three main stages in any defect assessment:

- Non-destructive inspection (carried out both prior to a component entering service and in service)
- Materials characterisation
- Structural analysis and fracture mechanics assessment

Both the NESC-1 and BIMET programmes have demonstrated the importance of successfully integrating these steps in order to predict accurately a fracture event or provide a reliable assessment of the significance of a defect for structural integrity. Each of the programmes has provided comprehensive data on crack size and geometry and materials properties as input to structural analysis and fracture mechanics assessment. Detailed 3-D finite element modelling using these data, together with well-defined loading conditions, has produced accurate predictions of crack initiation events under complex conditions, with due account taken of numerous structural and associated influencing features – see Table 1. Engineering assessment methods, while designed to serve primarily as failure avoidance procedures, have nevertheless proved capable of accurate predictions, when guided by post-test physical insights and with handbook limit loads modified in the light of results from more specific finite element analyses. Such experience, as exemplified in the case of BIMET, is invaluable in helping to formulate advice for extending the applicability of EAMs to increasingly complex fracture problems, whilst avoiding undue conservatism within the context of failure avoidance.

Whilst both the NESC-1 and BIMET programmes have done much to elucidate the general problem of transferability in predicting the fracture behaviour of components and structures, further research in this area is still needed. A key task concerns quantifying the way in which differences in crack-tip constraint are expected to increase the fracture resistance of components compared with that of high-constraint laboratory-scale specimens. The latter are tested to

Table 1   Summary of structural features/analyses covered by NESC-1 and DIMET programmes

Structural Feature/Influencing Feature	NESC-1				BIMET			
	EAM	Code Assmnt.	FEA	Local Approach	EAM	Code Assmnt.	FEA	Local Approach
1. Crack geometry								
– Under-clad crack			✓					
– Through-clad crack			✓	✓				
– Inclined surface crack							✓	✓
2. Constraint								
– Shallow crack								
– Near-surface regions of crack front			✓	✓				
– Multiaxial loading				✓				
3. Material properties								
– Ferritic base metal	✓	✓	✓	✓			✓	✓
– Austenitic base metal					✓		✓	✓
– Buttering layer					✓		✓	✓
– Weld fusion layer					✓		✓	✓
– Heat affected zone			✓	✓			✓	✓
4. Fracture mode								
– Ductile	✓		✓	✓	✓		✓	✓
– Cleavage	✓	✓	✓	✓				
– Effective crack front length			✓	✓				
– Ductile-to-cleavage								
– Intergranular fracture								
5. Thermal stress	✓	✓	✓	✓				
6. Residual stress								
– Structural weld								
– Dissimilar metal weld					✓		✓	✓
– Cladding			✓	✓				
7. Warm prestressing			✓					
8. Crack arrest								
9. Cladding effects								
– Effects on tensile properties			✓					
– Effects on ductile/cleavage toughness			✓					
– Effects on shallow crack initiation								
10. Dissimilar metal weld					✓		✓	✓

generate data for the reference toughness curves used in fracture assessments. While the so-called constraint effect is well known [9], its effects on the fracture behaviour of RPV components, where crack geometry and loading are generally quite different compared with test specimens, have not yet been sufficiently well validated as to justify acceptance by safety authorities. As part of the process of gaining this acceptance, it has to be demonstrated that any 'constraint benefit' for the fracture resistance of an RPV component at start of life, would not be eroded by deleterious changes in flow and fracture properties caused by in-service neutron irradiation. In the case of cleavage fracture, an associated issue concerns the question of effective crack front length in assessing the likelihood of brittle failure. Fracture toughness reference curves are obtained by testing specimens with crack front lengths in the order of 25 mm, whereas the length of crack front under consideration in a component assessment will generally be quite different to this. Weakest link statistics suggests a correction to the cleavage toughness reference curve of the form $(L_0/L_{eff})^{1/4}$, where L_0 is the reference length (usually 25 mm) and L_{eff} is the effective crack front length in the component being assessed. Two practical situations where clarification is needed in establishing L_{eff} relative to the physical crack front length, L, are (i) the case of long cracks, where $L >> L_0$; and (ii) where the crack driving force and crack-tip constraint vary significantly around the crack front. In each case, it seems reasonable to suppose that $L_{eff} < L$. In both cases, specific guidance is needed on the ratio $L_{eff}/L \leq 1$ that can safely be used under particular circumstances.

The NESC-1 test provided some evidence of the effects of crack-tip constraint and effective crack front length on cleavage.[4] The cleavage event that occurred at one end of Defect R immediately below the clad HAZ was shown to be consistent with local conditions of increased crack-tip constraint. (The absence of cleavage at the corresponding position on the other side of Defect R was attributed to the assessed higher toughness of material in that region.) Also, it was considered that an increase in effective crack front length from 25 mm to 200 mm is one of the factors that could help to explain the pockets of cleavage fracture observed at the deepest point of Defect R.

In conclusion, despite its overall complexity, a simplification in methodology for dealing with aspects of the transferability problem is beginning to emerge in certain areas. This is a development that has largely been driven by the numerous analytical insights that have resulted from comparing the predictions and outcomes of well-conceived structural features tests. As demonstrated in Table 1, NESC-1 and BIMET represent landmark programmes in this respect. Although conceived to address issues important to the nuclear industry, they have nevertheless helped in the development and validation of fracture mechanics methodology more generally. Here, non-nuclear industries may be expected to benefit, primarily through the development of simplified procedures with the proven capability of addressing specific transferability issues – e.g. fracture assessments of DMWs are commonly needed in non-nuclear energy-producing and process industries. It is important that this trend is maintained, with continued multi-disciplinary, international collaboration considered as being the most productive and cost-effective way forward. Follow-on programmes to both NESC-1 and BIMET are helping to achieve this objective.[10–12]

ACKNOWLEDGEMENTS

The author wishes to acknowledge the numerous colleagues who contributed to the execution, analysis and documentation of the large-scale experiments described in this chapter. He also acknowledges the support of the U.K. Health and Safety Executive and DG Research of the European Commission.

References

1. WCAP, 2000, 'Integrity evaluation for future operation of Virgil C Summer nuclear plant reactor vessel to pipe weld regions', WCAP-15617.
2. J. Varley, 'Networking for improved structural integrity assessment', *Nucl. Eng. Int.*, 1993, **38**.
3. J. B. Wintle, B. Hemsworth and R. C. Hurst, 'NESC: The Network for Evaluating Steel Components', *Proc. ASME-JSME 4th Int. Conf. on Nucl. Eng.*, Book 1389A1, 1996.
4. B. R. Bass, J. B. Wintle, R. C. Hurst and N. Taylor, 'NESC-1 project overview', European Commission, Directorate General Joint Research Centre Petten, EUR 19051 EN, 2001.
5. D. P. G. Lidbury, B. R. Bass, S. Bhandari and A. H. Sherry, 'Key features arising from structural analysis of the NESC-1 PTS benchmark experiment', *Int. Jnl. Press. Vessels & Piping*, 2001, **78**, 225–236.
6. D. P. G. Lidbury, A. H. Sherry, D. W. Beardsmore and R. A. Ainsworth, 'BIMET test 01: R6 assessment of a cracked bimetallic assembly with mismatch and weld residual stress', ASME Pressure Vessels and Piping Conference, Atlanta, vol. 412, 2000.
7. A. H. Sherry, D. P. G. Lidbury and D. W. Beardsmore, 'Influence of residual stress on ductile crack initiation and growth in dissimilar metal weld joints', 10th International Conference on Fracture, ICF10, Hawaii, 2001.
8. C. Faidy, Structural integrity of bi-metallic components (BIMET), Final summary report, Euratom Research Framework Programme 19941998, Nuclear Fission Safety, 2002.

9. A. R. Dowling and D. P. G. Lidbury, 'Local Approach modelling of constraint contributions to the ductile to brittle transition', Chapter 7 in *Plastic Flow and Structural Integrity, Proceedings of the 7th Symposium organised by the Technical Advisory Group on Structural Integrity of Nuclear Plant*, P. B. Hirsch and D. P. G. Lidbury eds, IOM Communications Ltd, 2000.

10. B. R. Bass, W. J. McAfee, P. T. Williams, D. J. Swan, N. Taylor, K. Nilsson, P. Minnebo, NESC-IV project: An interim report, European Commission, Directorate-General Joint Research Centre Report No. NESCDOC MAN (02) 04, Petten, 2002.

11. B. Eriksen, N. Taylor and F. Hukelmann, NESC-III: Developing a benchmark for structural integrity assessment of dissimilar welds in LWR piping, 3rd Int. Conf. on NDE in Relation to Structural Integrity for Nuclear and Pressurised Components, Seville, 2001.

12. D. P. G. Lidbury, B. R. Bass, S. Bhandari, D. Connors, U. Eisele, E. Keim, H. Keinanen, S. Marie, G. Nagel, N. Taylor and Y. Wadier, 'Validation of constraint based methodology in structural integrity: VOCALIST – overview', *Proc. ASME Pressure Vessels and Piping Conf., Atlanta*, vol. 423, pp. 83–91, 2001.

CHAPTER 7

Structural Integrity in Aero Gas Turbine Engines

R. S. J. Corran, S. J. Garwood and A. R. M. Walker

Rolls-Royce, plc, PO Box 31, Derby, DE24 8BJ, UK

ABSTRACT

In Western Europe, the Joint Aviation Authorities, a consortium of aviation authorities from participating member states, publish requirements covering all aspects of aviation. The part which has specific application to aero engines is JAR-E. The rules given in JAR-E set standards of strength and margins which must be met by all power plant for civil aviation usage.

The main concerns for structural integrity in aero gas turbine engines are identified as behaviour as a result of ingested objects, low cyclic fatigue failures, crack growth and coalescence of cracks. A failure modes and effects analysis identifies those components in the engine whose failure may lead directly or indirectly to hazardous effects. It is necessary to demonstrate that the duty of such Critical components and the variation in their strength combine in such a way that failures cannot occur. The approaches used to establish the cyclic capability of such components are described.

With other components, notably blades, failure is designed against by providing a containment shell around the rotating stage which will prevent the release of a single blade outside the engine nacelle. As well as this, the engine is tested to demonstrate robustness in operation for the hot sections including combustors and turbine blades.

With fan blades, however, which are the first stage and therefore more likely to be subjected to ingested object damage, it is necessary to show that they can sustain impacts by small and medium birds but still retain part or all of their thrust capability. For impact with large birds, however, it is simply enough to show that the engine can be shut down safely. In the event of a fan blade becoming detached, the imbalance causes large structural loads on the engine and engine mounts. A test is generally required in which a blade is released at the most arduous point in the engine operation and both the containment system and the ability to shut the engine down safely is demonstrated.

Engine mounts, which connect the engine to the airframe, may be designed with damage tolerance where a crack is assumed in the most highly stressed location and, where applicable, another is assumed in the secondary load path. In such cases the inspection interval is related to the number of flight cycles to burst from the specified crack sizes.

Casings and combustors are two examples of thin shells used in gas turbine engines and their structural integrity is also described.

1 INTRODUCTION

The modern gas turbine is used for powering aeroplanes and ships, generating electricity and pumping oil and gas. It has an extremely high power density and it is no surprise, therefore, to find that maintaining structural integrity presents a number of challenges. The industrial Trent Power Unit, for example, can raise over 50 MW of electricity but is still small enough (just!) to be carried on the roads by a low-loader truck. A gas turbine consists of a number of concentric spools, these being shafts connecting together *compressor* stages at the front to *turbine* stages at the rear, a stage consisting of a single row of rotating blades and stationary vanes. Air sucked into the compressor is pressurised so that when entering the combustion chamber the flame and consequent hot gas is directed towards the exit nozzle through the turbine. The turbine stages extract power for driving the compressor, At the front of the aero gas turbine engine, the first compressor stage, called the *fan*, provides not only the initial pressure increase to air passing into the core of the engine, but also, due to the oversize nature of the blades, flow and thrust through the bypass duct. The hot gas exiting from the *combustor* of a modern three spool engine for use on large wide-bodied aircraft, e.g. Trent 700, is at a temperature in excess of 1700K, hot enough to melt the turbine blade if the cooling-air fails, while the inlet temperature is close to ambient. It is not surprising, therefore, to find that thermal gradients provide a large part of the loading experienced by many engine structures. The spools rotate at high speeds relative to their diameter. The high pressure spool, diameter about 650 mm, for example, rotates above 10 000 rpm while the low pressure spool, which contains the fan, is of much larger diameter, over 2 m., and rotates more slowly, at about 3300 rpm.

The major structural integrity issues with safety implications are as follows:

1. *Ingested objects* leading to damage to rotating blades and stationary vanes. The objects may be *hard*, such as pebbles and stones sucked up off the runway, water in the form of ice or hailstones or *soft*, such as birds. As the bird gets bigger the probability of ingesting more than one becomes less, and the rules for certification recognise this by requiring the engine to maintain a percentage of thrust (75%) for a flock of birds whose number is determined by the intake area of the engine, there being more small (0.085 kg) than medium birds (0.68 kg). For a single large bird, currently 4 lb (1.8 kg), the engine must not release any high energy fragments which could hazard the airframe and must also be safely closed down. For a large ingested bird, a blade may be highly damaged leading to the potential for immediate release while at the other extreme of a small hard object, the resulting impact on the blade can lead to a sharp notch like feature which

acts as a site for initiating a fatigue crack. The fatigue may be high cycle, low cycle or a combination of the two. It must be demonstrated by test that the debris would not hazard the airframe.

2. *Low cycle fatigue* failures of specific parts, which may be rotating, e.g. discs and spacers, or static, e.g. engine mounts and combustor casings. The released energy at disc burst is so large that it is not feasible to contain it. Consequently failure of such parts leads to high energy fragments which escape the engine nacelle and may hazard the airframe. For static parts the concern is less to do with released high energy fragments but more with the inability to control the engine and aircraft. The combustor casing, for example, ensures that the hot gases of the combustion process do not act to soften key engine parts and even the wing.

3. *Crack growth and coalescence of cracks* in certain parts of the combustor could lead to misdirection of the combustion flames which then play upon key areas of engine and airframe structures leading to a hazardous condition.

The objective of this paper is to explain the different approaches taken to ensure the structural integrity of a range of engine parts. For civil aviation applications the requirements for structural integrity are embodied in regulations.[1,2] In

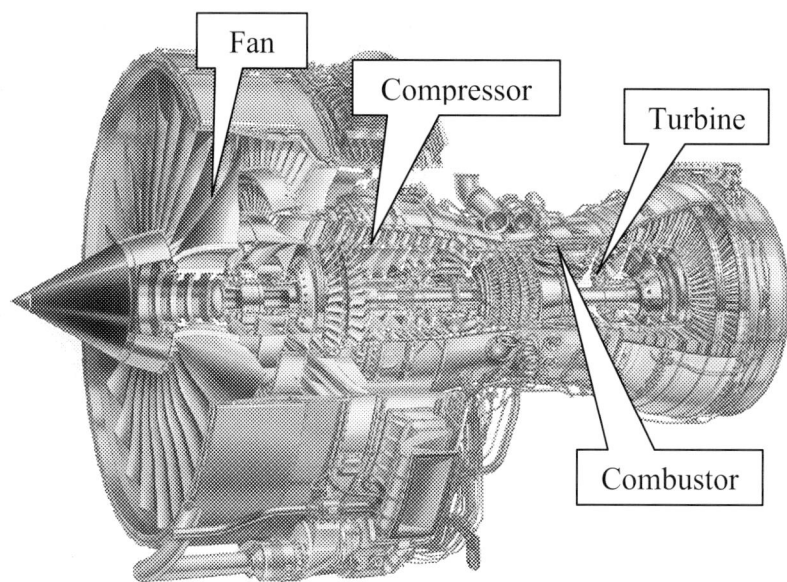

Fig. 1 The Rolls-Royce Trent 700 engine.

the USA, part 33 of the Federal Aviation Regulations (FAR) is relevant to the certification of the engine while in Europe the Joint Aviation Regulations Engines (JAR–E) are used. These requirements are a combination of *absolute rules* where the Original Equipment Manufacturer (OEM) must show that certain types of failure will not occur, and *pragmatic rules*, designed to ensure a level of integrity in the engine which, while not showing that failure can never occur, has been found to be acceptable.

2 CLASSIFICATION OF PARTS

At the design of a new engine, a Failure Modes and Effects Analysis (FMEA) is performed as required by the regulations JAR-E[1] and FAR 33.75.[2] This is a systematic examination of the engine behaviour under postulated normal and accidental conditions. Initially each part in turn is assumed to have failed and the consequences to safety are assessed. Hazardous effects are defined which include the release of high energy fragments and the inability to maintain control of the aircraft. Those parts whose single failure may lead to hazardous effects and which have to maintain a high level of integrity are defined as CRITICAL in JAR–E. Such parts are subject to special controls to ensure that they meet a particularly high standard of manufacture and that they are capable of meeting the duty required. These controls extend not only to melting, forging, machining and inspections but also to the close definition of the duty in terms of temperature and stress and to an active management plan in service.

In order to maintain the high levels of integrity required, the material supply is closely controlled. The specification of titanium alloys, for example, covers not only acceptable melt methods and the necessary inspections both in billet and forging shape but also the source materials to be used in the melt, preventing the use of certain forms of the raw material which have been found to give unacceptable rates of melt anomalies. In order to achieve a particularly high standard of inspection of the volume of the disc forging, an intermediate oversized shape is used which is optimised for inspection allowing full coverage from a number of different directions and employing both normal-to-surface and angle scans.

A second task of the FMEA is to identify when hazardous effects can arise from a concatenation of events. Thus, for example, a failure to disengage the starter motor can lead to a backdrive event which could burst the disc on the starter motor in overspeed. Such a hazardous effect would be caused by two events, the first being the failure to disengage while the second would be the failure of the starter in the consequent overspeed. Where it can be shown that the probability of such a hazardous effect is Extremely Remote (taken in this context as being less than one such event per 10^8 hours of operation of the engine), then it is permitted to regard this as so low as not to need further consideration at the design phase. Where the probability is higher than Extremely Remote, the designer must take further action, such as providing containment for the starter motor in the example above, to ensure the integrity of the engine.

While provision is made through the classification of parts as critical for cases where the designer is unable to provide for the failure of a single part without the possibility of causing hazardous effects, it is nevertheless the duty of the designer to minimise the number of critical parts and the ways in which hazardous effects can arise. So, for example, the discs in the turbine which drive the compressor blades will overspeed if the torque carrying shaft fails. The designer must then ensure that the disc design is capable of sustaining the level of overspeed which is predicted in this event, thus removing this mode of failure from the list of hazardous events. Although the shaft may not be critical, the disc remains critical because of the inability to contain a burst disc.

3 CRITICAL ROTATING PARTS

The main integrity issues for critical rotating parts are discussed below. Nearly all such parts, which encompass discs, spacers and, to a lesser extent, shafts, are made through a conventional cast and wrought route by which the melted metal is used to make an ingot, which is then converted into a billet by work to reduce the diameter and increase the length. The billet is split into 'uses', also known as 'mults' in the USA, each of which is forged to a shape which encloses the final disc shape. This 'black' forging is machined to an *intermediate shape* for ultrasonic inspection of the volume after which it is machined to the final shape. Inspections include ultrasonic examination of the billet and intermediate shape, and etch and penetrant inspections of the intermediate shape and final shape.

In the last two decades increasing use has been made of powder metallurgy, in particular to produce discs for high temperature operations such as the high pressure turbine. Here a molten metal stream is atomised and the resultant powder sieved to ensure no particles bigger than the sieve size, typically 50–150 μm. The powder is hot isostatically pressed into a 'log', this being a cylindrical bar, and then forged into a shape enclosing the final shape of the disc. From this point onwards the process and inspections techniques are identical those used for conventional cast-and-wrought forged discs, although the ultrasonic inspection standard achieved may be better due to the uniformly smaller grain size and hence inherently lower 'noise' level.

Table 1 Manufacturing of a drum or assembly with sources of anomalies and inspections. Note that the inspection after any step does not only focus on issues arising in that step.

Step	Product	Typical issues	Inspections
Melting	Ingot	Inclusions Hard Alpha Contamination Segregates	Chemistry
Conversion	Forged Billet	Strain induced porosity (in Titanium alloys)	Ultrasonic Inspections Etch on slices Mechanical properties
Forging	Rectilinear machined forgings	Laps / Folds Structures Cracks Strain induced porosity (in Titanium alloys)	Ultrasonic inspections Etch + fluorescent penetrant inspection Visual Mechanical properties and structures
Machining	Finish machined disc	Damage Microstructure Cracks	Etch + fluorescent penetrant inspection Visual + binocular
Welding and post weld heat treatment (PWHT)	Drum or Assembly	Pores and cavities Cracks Lack of fusion etc.	Etch + fluorescent Penetrant inspection Visual + binocular X-ray

3.1 DESIGN AGAINST OVERSPEED

At the simplest level, design against overspeed failure is demonstrated by spinning a disc up to the required speed at the most adverse engine conditions in terms of temperature at which such an overspeed could be considered possible. In undertaking this test there is recognition of the fact that the disc tested is not necessarily a minimum disc and that the speeds are for a deteriorated engine rather than for one just overhauled. In practice use has been made of correlation of the overspeed capability of a disc to both its tensile properties (proof and ultimate strength) and the distribution of hoop and radial stress in the disc. Using such correlation it is often possible to design a disc on the basis of previous in-house experience of material properties and projected service duty and avoid the need for further spin testing for certification by the regulatory authority.

3.2 FAILURE FROM FATIGUE DUE TO CYCLIC USAGE

For over 30 years, Rolls-Royce has based all cyclic lives of discs on the results of component tests, normally achieved by spinning at a temperature representative of engine conditions at a speed to develop a peak stress greater than that in the engine flight cycle. The life calculation is based on features in the disc, these being, generally, the *rim* including blade attachment slot (firtree or dovetail), the *diaphragm* or *web*, the *bore* and *cob* (hub) and the *drive arm*, to which can be added other stress raising features such as holes. Each feature is assessed individually and the life of the disc is the shortest life calculated.

In the simplest method, known as the *Safe Life* approach, a disc is tested at a stress no less than that in service and a temperature no less than the service temperature at which the peak stress occurs and a life to 'first engineering crack'. The engineering crack is the size of crack that can be reliably detected using conventional inspection methods, conventionally 0.75mm surface length. A fraction of the cyclic life thus established is declared as a 'safe-life' for engine operation, duly factored for engine stresses, temperatures and the effect of minor stress excursions in the engine flight cycle. The general assumption is that the ratio in *life* between a *best* component ($+3\sigma$) and a *worst* component (-3σ) tested under identical conditions is 6:1 from which a ratio between the 95% best component and the assumed 'worst' can be deduced as almost exactly 4:1, (see Fig. 2). The lives are assumed to conform to a log normal distribution in which six standard deviations are equivalent to a ratio of 6. Hence for a 95% component at $+1.645\sigma$, the ratio to the assumed 'worst' component is given by $(3 + 1.645)\sigma$ or $6^{(4.645/6)}4.003$. Assuming that the component tested is a 95% best component and declaring the life at the worst level gives a 1 in 20 chance that the life is such that 1 in 741 components will exhibit small amounts of cracking in service at the full life.

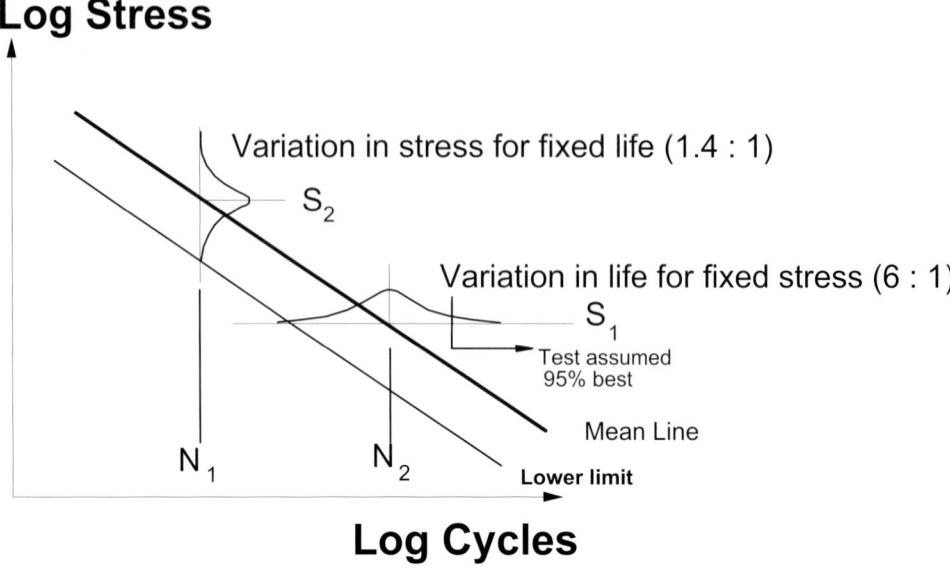

Fig. 2   The use of scatter factors in declaring a safe cyclic life in the engine

Clearly testing more than one component gives a higher confidence that the sample mean is closer to the population mean and while still assuming that a 95% best sample has been tested it is possible to reduce the safety margin.

The establishment of an equivalent stress variation is useful since it is customary to test at a stress somewhat higher than that in the engine cycle. For example a test at 30% overstress would be equivalent to a 4 times longer life at the datum stress level. This allows the test duration to be shorter and the use of previous spin test results performed at a higher peak stress than that in the engine cycle.

Since the early 1980s Rolls-Royce has employed a *DataBank* approach in which a number of disc features are combined into a common correlation. Use of this approach better exploits the number of discs tested, incorporating a range of features, and allows the inclusion of specimen test results. This expands the range of conditions tested and can allow testing of conditions which arise in transient conditions in the engine but which are not easily achievable in steady state on a component test. The DataBank correlation is currently based on a fracture mechanics assessment in which an effective initial flaw size (EIFS) is postulated which, when grown under conventional linear elastic fracture mechanics (LEFM) long crack 'Paris' modelling, would have given the life actually exhibited by the component or specimen at the relevant test condition. A statistical treatment of the EIFS values allows the identification of an extreme which represents the 'worst damage' arising from a well controlled manufacturing route. Calculating the burst life at each feature in the disc using this extreme EIFS takes account of the peak stress and stress gradient. The requirement that all test discs are controlled to have surface conditions consistent with the engine production disc ensures that the important effects of surface condition are also captured. The life declared is twp-thirds of the burst life using a 1 in 1000 worst EIFS value.

In both *DataBanks* and the *Safe Life* approach, a relatively tight scatter in disc lives is integral with the method of life declaration. The narrow scatter exhibited by the disc when manufactured either by the cast and wrought route or the powder metallurgy route means that lives declared are significantly greater than the life-to-burst of cracks which could be just missed by the various inspections, both sub-surface and surface, which are applied during the manufacture of the disc. The integrity of the disc is therefore dependent on the careful choice of, and control of, processes applied to the melting, forging and manufacture of the discs and also to their handling in service to ensure that the scatter in the fatigue lives of the population is as tight as is assumed. Primarily this means that anomalies, such as arise in melt (inclusions, contamination, adverse microstructure such as hard alpha phase titanium), during forging (laps, strain induced porosity), in manufacture (abused microstructure, scratching, scoring) and in maintenance (scratching, scoring) must be either eliminated or kept at a very low level both in frequency of occurrence and size. In practice the majority of anomalies are not cracks and will take a certain portion of the cyclic life to develop into a propagating crack. Furthermore, although the reliable penetrant detection capability is around 0.75 mm for full field methods not especially focused onto a specific feature or area, this does not mean that there is no detection capability at sizes lower than this. In practice outbreaks of inclusions and other such anomalies will be found because some if not all will be visible from the inspections. Hence the

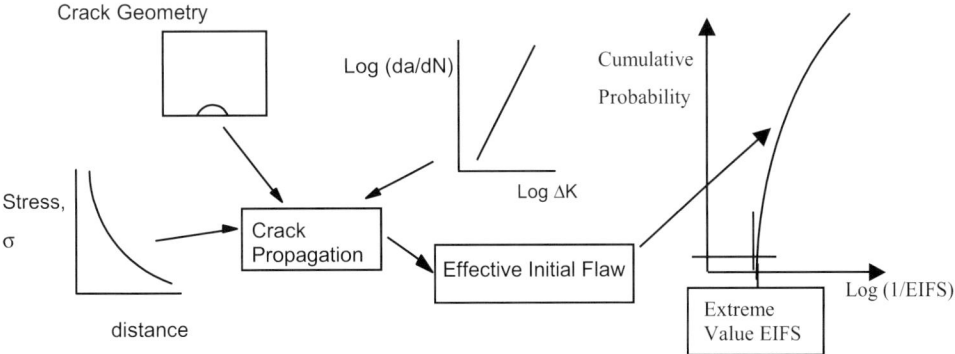

Fig. 3 Fracture Mechanics correlation as used in the Rolls-Royce DataBank lifing method

process controls consist not merely of inspections but must also extend to procedures for sentencing material adjacent to finds which can be assumed 'guilty by association'. Structural integrity from these procedures can be substantiated by the low occurrence of burst discs from these causes in the past 30 years. While a few instances of lapses in process control have occurred in the industry, the procedures are constantly reviewed to enforce the tight scatter inherent in the lifing methods.

3.3 MAINTENANCE OF SHAPE UNDER ELEVATED TEMPERATURES

The purpose of the disc is to maintain a row of blades rotating in the gas stream at a fixed radial and axial position in the engine. The choice of material for the disc is much affected by this requirement especially in the hotter parts of the engine. In the pursuit of ever lower costs, the designer will seek to extend the use of lower-temperature capable, and therefore cheaper, materials as far into the hotter parts of the engine as possible, consistent with the requirement to limit the amount of cooling air in the interests of cycle efficiency.

Design against creep deformation is achieved mainly by ensuring that the averaged stresses are below those causing an acceptable maximum amount of *creep deformation* during the life of the engine. Generally it is the radial deform-ation of the disc which is of concern since this causes the blades to rub against the outer annular limit to the gas path. Maintaining such tip clearances to a necessary minimum amount plays an important part in achieving an economically accept-able performance for the engine and this is achieved by using abradable linings to the outer annulus casing. Nevertheless excessive radial growth would allow the blades to receive such a heavy rub that multiple blades could become detached and would then fly out radially, which could hazard the airframe. It is common practice to design the disc and blades such that a single release does not cause other blades to be released. With these restrictions it is only necessary to show that a single blade will be contained since multiple blade release does not occur. Should it not be possible to demonstrate that multiple blade release cannot occur, then the containment system must be designed for as many blade releases as an FMEA shows can occur.

An average stress across the diaphragm (web) is used as the criteria for limiting creep deformation because creep is regarded as a bulk behaviour of the material, not highly sensitive to surface condition, leading to a redistribution of stresses towards a more uniform state. Of course a more complex analysis using a finite element model is possible and this would capture the biaxial effects in the creep deformation, important where the creep behaviour shows sensitivity to both the maximum principal stress and the von Mises' stress. In general, the restriction of deformation in creep is sufficient to ensure rupture does not occur.

4 BLADES AND VANES

Engine blades and vanes are subject to both low cycle fatigue loads (LCF) associated with the flight cycle and high cycle fatigue (HCF) related to vibrations at the natural frequency of the blade or vane. While LCF loads can be analysed for stress levels, stresses due to vibration have traditionally been established by modal analysis and engine measurements, although better predictive methods are now becoming available. The amplitudes of the HCF are limited by the damping and other non-linear behaviour of the blade. The blades rotate at high speeds in the gas path, and ingested objects, which vary in type, shape, size and velocity and impact point, may cause damage ranging from erosion, corrosion, nicks and dents through to significant plastic deformation (cupping). While the designer seeks to ensure that a damaged blade or vane can resist the engine loads, their fatigue lives are not as predictable as for discs, and less reliance on their continued structural integrity is assumed. The engine designer is therefore required to provide containment for a single blade release from any stage in the engine. For a fan blade, with a mass of about 10 kg in a large civil engine, containment is a major issue. As the blades get smaller towards the back of the compressor containment is less difficult, while turbine blades lie somewhere between.

4.1 FAN BLADES (see Fig. 4)

Fan blades must withstand soft body impact (birds) as well as sustaining hard body impacts which damage the surfaces of the blade. Both soft and hard body impacts are called foreign object damage, FOD. For small and medium birds the issue is to design the blades in such a way that the 'cup', a region of plastic deformation at the leading edge caused by the bird impact, does not deteriorate the engine performance such that continued safe flight and landing of the aircraft is imperilled. This capability is demonstrated by an engine test, analysis being used to determine whether the medium size, with fewer birds, or the small size, with more birds, is most critical. The engine is run to the defined most arduous condition and the birds, or artificial birds, are fired into the intake using a gas gun to simulate the forward velocity of the aircraft. Although an initial momentary lowering of thrust is allowed, the requirement is for the engine to settle onto a sustained level of no less than 75% of the nominal level.

When impacted by a large bird, currently defined as 1.8 kg in the regulations, but soon to be 3.7 kg, it is required to demonstrate by test that a blade is not released and that the engine can be shut down safely. Because of the mass of the bird, the blade can be expected to undergo large structural displacements due to the impact. The issue, initially, is to ensure that the blade remains intact and does not detach, either as a whole body at the root or just the tip portion. The

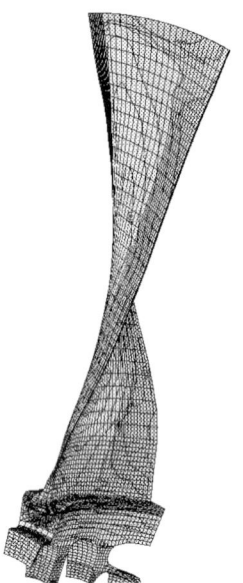

Fig. 4   Engine test (Trent 800) and analysis model (V2500) of fan blades.

capability to sustain such an impact is demonstrated by test, the engine being again run to a condition defined by analysis as being the most arduous for impact and the 'bird' being fired-in using a gas gun. For the large bird where the requirement is not one of the engine maintaining thrust, there is more possibility of combining analysis with a test of just the fan stage to demonstrate that no hazardous condition results to the whole engine. The issues about engine handling and structural response will be dealt with below when dealing with blade containment.

The fan blade is also subject to low cycle fatigue from loads associated with flight cycles (centripetal forces, gas pressure differential) and high cycle fatigue from excitation such as flutter and resonance. Smaller fan blades are single piece forgings similar to discs, but with more aligned grain structure in the radial direction and different surface conditions. Larger fan blades are fabricated, generally using some form of diffusion bonded superplastically formed Titanium 6Al/4V to give a light but high stiffness structure. In the latter case the stress analysis is more complicated, but for both types of blade the LCF life is demonstrated by testing a blade, or set of blades, in a cyclic spin test. Because it is necessary to use an engineering vacuum compared to operating pressures ranging from normal atmospheric, $\sim 100\,\text{kPa}$, at take-off to $26\,\text{kPa}$ at $10\,000\,\text{m}$, the speed must be adjusted to yield representative stresses. Furthermore it is not possible to superimpose the flow-induced vibrations since there is little gas flow. Derivation of a 'safe-life' follows procedures essentially identical to those for discs, although now it is necessary to allow for the increased number of blades (typically 22–26 per set) as opposed to the single disc in the statistics.

From an aerodynamic point of view, a smooth surface to the blade with extremely sharp leading and trailing edges is ideal, giving the best pressure rise for the minimum work. From the integrity point of view a deep surface compressive stress, even at the expense of surface smoothness, combined with a blunter leading edge is likely to be more damage tolerant.

The main concern with hard body impact damage is that it will act as an initiation site for cracks which are grown by HCF. In low cycle fatigue, related to one major cycle per flight, there should be ample opportunity to detect the damage during ground inspection and hence remove the blade after the FOD event but before failure. However in high cycle fatigue, given the relatively high frequency of vibration of the blade, from around 70 Hz upwards depending on size and mode, the accumulation of cycles can be quite rapid which could lead to failure within a few flights once a crack is long enough to be sensitive to the vibration stresses. To ensure failure does not occur vibration amplitudes must be maintained at sufficiently low levels such that damage is not accelerated to burst at high frequencies. An HCF test of the blade is performed by forcing it to vibrate at a resonant frequency and the amplitude controlled to a level related to the engine requirement. The objective here is to establish that at the level of HCF experienced in the engine, the low cycle fatigue life is not significantly affected thus giving an acceptable life between inspection intervals.

4.2 COMPRESSOR BLADES (see Fig. 5)

Most of the issues surrounding fan blades also apply to compressor blades. Because the initial impact of ingested birds will be on the fan, compressor stages are more likely to experience a 'softer' impact in which the bird is more cut up and liquefied. Since the bird tests are on the engine, then the performance issues for compressor blades will be tested at the same time. Hard body impact (FOD) is demonstrated by testing, in HCF, blades which have had a notch machined representing the damage arising from such impacts. Low cycle fatigue is not generally tested by spin test, but a substantial number of cycles is often accumulated in development testing. Furthermore the engine is put through vibration searches in which it is cycled in speed between minimum and maximum operating speeds as it deteriorates during endurance testing specifically to find speeds at which resonance occur. Post-test inspection is then the means of demonstrating the integrity of the baldes.

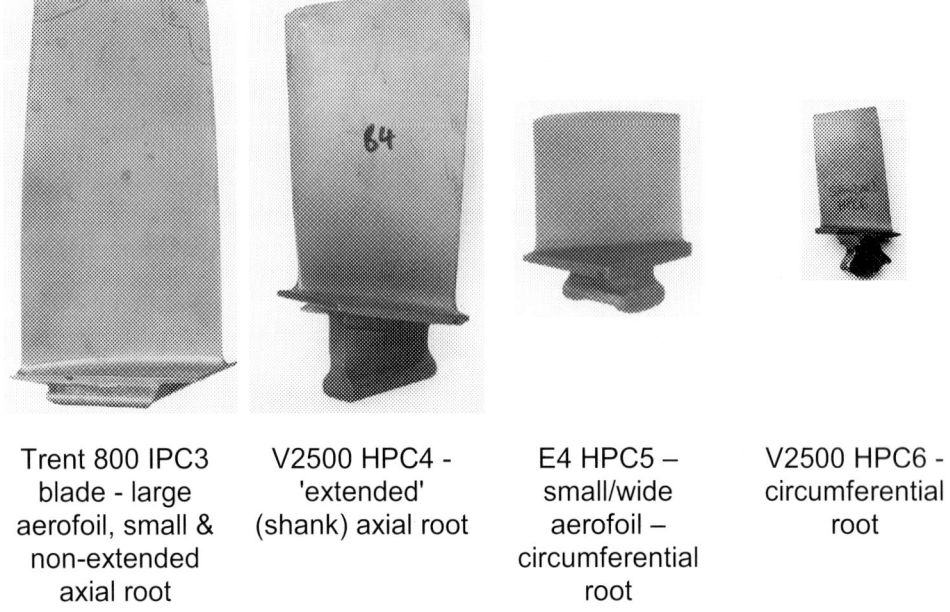

| Trent 800 IPC3 blade - large aerofoil, small & non-extended axial root | V2500 HPC4 - 'extended' (shank) axial root | E4 HPC5 – small/wide aerofoil – circumferential root | V2500 HPC6 - circumferential root |

Fig. 5   Compressor blades (not to scale).

4.3 TURBINE BLADES

Turbine blades are subject to the hot gas on exit from the combustor, hotter than the melting point of the blade material. Such blades are therefore cooled, and have four principal modes of failure:

- Creep rupture
- Fatigue
- Chemical degradation from the hot gas stream, mainly oxidation but may include corrosion from combustor products
- High cycle fatigue

Any blockage to the cooling holes is critical to the operation of the blades because, as noted earlier, the hot gas stream is generally above the melting point of the blade material and loss of coolant results in either severe blade damage or complete loss of the blade.

The main vehicle for showing the durability of the turbine blades is the *type test*. This is an extended cyclic running of the engine designed to expose any issues of resonance through the vibration searches mentioned earlier, and to subject the blades to a combination of temperature and stress which are on the more arduous

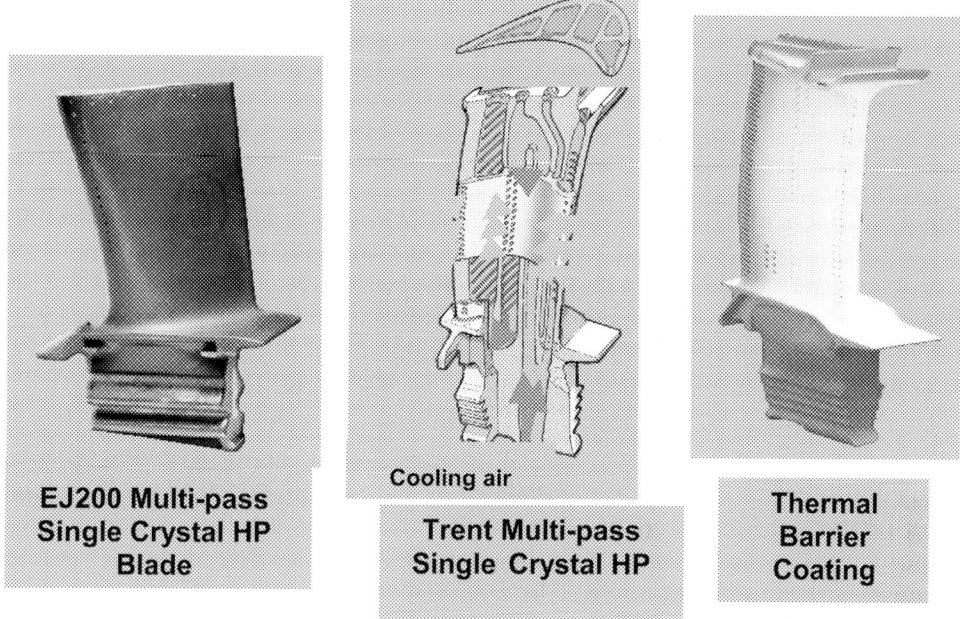

Fig. 6 A selection of turbine blades designed by Rolls-Royce.

side of normal operation. Given the primary nature of centrifugal loading on the blade and that it is operating in a hot gas stream, creep is clearly a major design consideration. Since it is a certification requirement to contain single or likely multiples of blades in the event of their failure, there is no immediate safety issue in isolated blade failures. However should the life of the blade be so low that simultaneous failure of more than one aircraft engine in the same flight becomes a finite possibility, then clearly there must be a safety implication. For twin engined aircraft there are limits on the flying time an aircraft can be away from a diversion airport, known as the **Extended Twin Operations** (ETOPS). The objective of the type test is to demonstrate that there is no possibility of extremely low life failures which could lead to in-flight engine shutdowns. The regulations call for an 150 hour type test.

5 OTHER STRUCTURAL PARTS

The engine static structure serves a number of key purposes. It locates the bearings for the engine spools in space and hence transfers the thrust from the spools to the airframe via the engine mounts. It must do this with minimal distortion especially on the static outer surfaces of the gas path which, by changing the blade tip clearances, has a direct effect on engine efficiency. Again, given the loads and powers involved in the engine, the structure has an arduous duty to fulfil.

5.1 MOUNTS (see Fig. 7)

Engine mounts which interface with the airframe are subject to the airframe regulations, JAR 25,[3] rather than the engine regulations, JAR–E. Structural integrity is often demonstrated using a damage tolerance approach rather than the safe life approach used for discs. The engine mount may have dual load paths such that they become fail-safe. The damage tolerance requirement is split between the primary path, in which a larger flaw is assumed, and the secondary path, in which a smaller flaw is assumed. The structural integrity is established by showing that the life for the given flaw size to grow to failure of the primary path is a specified number of inspection intervals while the secondary path is able to maintain a further number of inspection intervals before its given flaw size grows to failure. Typically the primary flaw size is assumed to be a 1.25 mm. corner crack in the most damaging location while the secondary flaw size is 0.125 mm. also in the most damaging location.

Loading on the engine mount is more complex than many other engine parts. The wing is subject to relatively large movements, both in taxi-ing due to low frequency vibration and in flight due to buffeting by the wind and this,

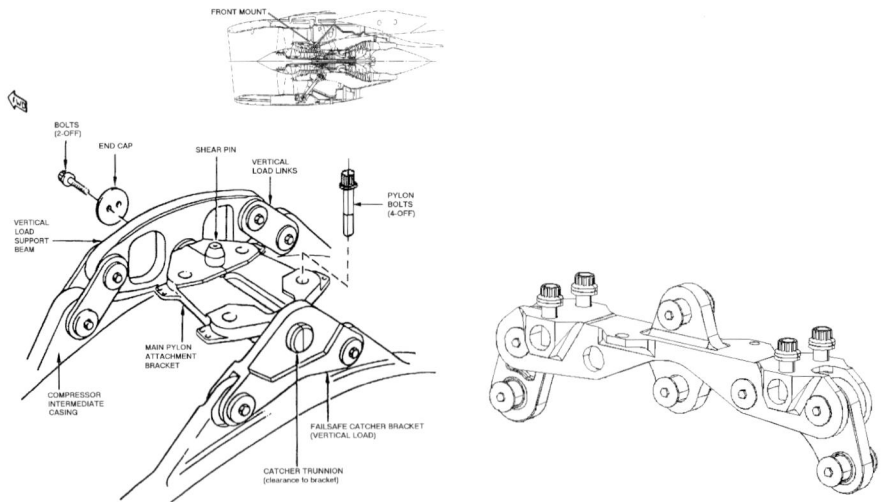

Fig. 7   Front and rear engine mounts for the Trent 700. In both mounts there is a dual load path such that failure of any linkage causes the secondary path to take the load.

superimposed on the changes in mount loading due to engine thrust (which are more predictable), provides a complex load spectrum for each flight. Based on surveys of wing loading and experience in previous airframes, the practice is to provide a range of amplitudes and frequencies of occurrence, the higher amplitudes receiving fewer occurrences in the life of the airframe. While analysis is used extensively to identify the most damaging feature of the mount systems, it is common also to require a test to demonstrate the mount's capability. This will involve the generation of a crack from a machined notch and also the reduction of a mix of highly complex flight cycles to a simpler spectrum which can be achieved in a sensible time and at sensible cost on a load frame. Analysis is used to translate the reduced cycle into equivalent full cycles.

5.2 PRESSURE CASINGS

Although the pressure in a gas turbine does not normally rise to remarkably high levels compared to, say, chemical reactor vessels, nevertheless the quantity of gas passing through the engine is substantial, about a tonne per second for a large high bypass ratio fan for a modern twin engine wide bodied airframe. The pressure casings also, as in other pressure vessels, fulfil an important containment function. The combustion chamber outer casing (CCOC), for example, maintains the pressure in the combustor and prevents the hot gases from impinging upon other important engine parts, see Fig. 8.

Under the JAR,[1] *critical* pressure casings must fulfil all the planning and integrity requirements of other *critical* parts. In addition they must achieve the same capability as for all other pressure parts, i.e. being capable of sustaining the maximum possible pressure without bursting and to endure the cyclic life under fatigue of normal working pressure. Here the pressures are defined as:[1]

Normal Working Pressure. The maximum pressure differential likely to occur on most flights including any pressure fluctuations as a result of the normal operation of valves, cocks, etc., where these could produce significant surge pressures.

Maximum possible pressure. The maximum pressure differential which could occur under the most adverse combination of operational conditions likely to be experienced in service, together with failure of any relevant parts of the engine or control system, or combinations of failures . . . .'

It is required to demonstrate a margin by ensuring no permanent distortion at:

   (i)   1.1 times the maximum working pressure or,
  (ii)   1.33 times the normal working pressure or,
 (iii)   35 kPa above the normal working pressure, and

no burst at:

(i) 1.15 times the maximum possible pressure or,
(ii) 1.5 times the maximum working pressure or,
(iii) 35 kPa above the maximum possible pressure.

These rules imply similar requirements to those embedded into pressure vessel codes, e.g. BSI PD 5500, the margins being almost identical.

5.3 COMBUSTOR PARTS (see Figs 8 and 9)

Combustor parts are designed to direct the gas around the flame such that metal surfaces are washed by cooler air from the exit of the compressor so that the hot combustion products do not impinge directly. Such parts are made from thin metal sheet with many holes to allow the flow of the cooling gas to maintain the film of cooler air against the metal wall. Temperature gradients can be severe. Primary stresses, defined as those in equilibrium with the primary pressure load, are low apart from in pressure casings. Cracking of non-pressure retaining combustor parts does not necessarily render them unfit for purpose since their function is directing the gas flow and small cracks cause little loss of this ability, not least

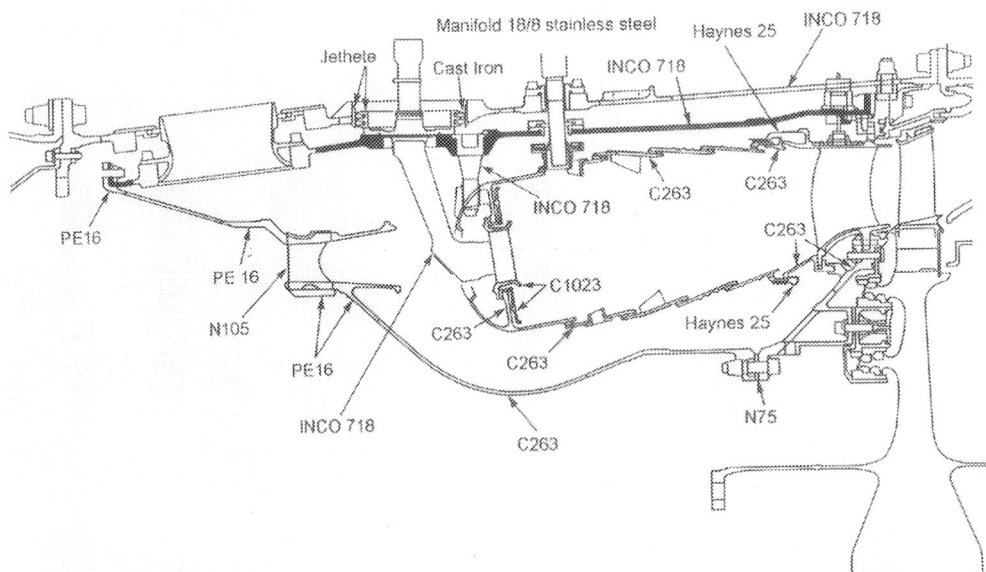

Fig. 8 The combustor in a modern 3 shaft engine showing materials used. The CCOC is the outermost Inco 718 shell.

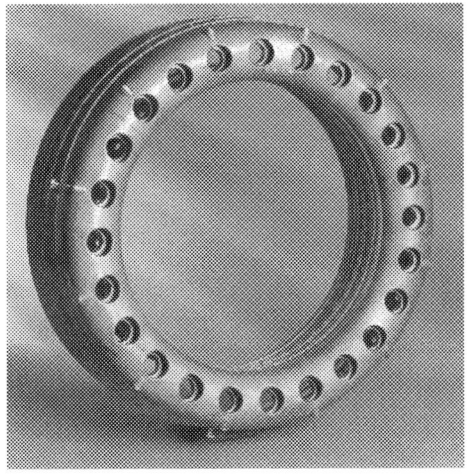

Fig. 9 The combustion ring from the Trent 500 engine.

since the combustor parts are already full of small perforations for the cooling air. Because of the severe thermal gradients, which can give rise to very high secondary stresses, some cracking is not uncommon. Integrity of the engine can be lost if sufficient cracking occurs, perhaps by coalescing of a number of smaller cracks into one large crack. When the cracking becomes extensive, parts of the combustor can distort in the pressure flow such that the flame is directed onto metal pressure casings leading to weakening, pressure burst and an uncontained fire. In such cases, therefore, integrity is an issue of preventing extensive crack growth.

5.4 RELEASED BLADE (see Fig. 10)

As has been discussed in section 4 above, fan blades are designed and demonstrated to survive specified sizes of bird ingestion. There is no certainty, however, that this will make the engine safe to operate because the requirement for bird-strike capability does not, and probably cannot, cover for all possible instances. For example, it is not, under current regulations, required to demonstrate either of the following scenarios:

1. Ingestion of more than one 1.8 kg bird. The increasing diameter of the intakes of the latest engines make this more possible and this is part of the reason for moving to a larger bird standard.
2. Ingestion of a bird bigger than 1.8 kg (or even 3.7 kg with the future requirement). Such birds do exist.

Fig. 10   An engine prior to blade-off testing. The release blade is coloured to assist in analysing the trajectory after release.

Because failure of the blade cannot be ruled out, it has been necessary to design the engine such that failure of a blade does not give rise to hazardous engine effects. The designer is therefore required to demonstrate:

> that any single compressor or turbine blade will be contained after failure and that no hazardous condition can arise to the aircraft as a result of other Engine damage likely to occur before Engine shut down following a blade failure.[1]

Note that compressor here includes both fan and compressor.

5.4.1 Blade-off loads

The fan blade is the largest blade in the engine and gives rise to a significant out-of-balance load when released. While not a regulatory requirement, JAR–E advisory material recommends that the most severe case of both compressor and turbine blade releases should be tested by releasing a blade at the most adverse engine condition. Generally this will be a fan blade in the compressor. The engine bearings, which both locate the shafts and transmit the thrust from the engines to the aircraft through the engine static structure and the mounts, must be designed to withstand the considerable transient forces which arise from the out-of-balance. In this test the control levers for the engine must not be moved for 15 seconds after the blade release to represent the response time likely in operation, although since the recommendation is that the test should be performed on a complete engine, an automatic shut-down by the engine itself in a shorter period is allowed.

While the regulations require the demonstration of the engine's capability to sustain the blade-off load at the most adverse conditions, it is not always immediately clear what these will be. In practice the engine designer will undertake a number of calculations to determine the 'worst' condition and test that, using the modelling methods developed to extend the demonstrated capability to other conditions, perhaps not so easily tested at ground level.

5.4.2 Blade containment

A large modern civil fan blade may have a mass in excess of 9 kg and rotates at around 3300 rpm at take-off. If released, such a large mass with high kinetic energy and momentum represents a hazard to the airframe. The engine designer provides a containment casing to prevent such a blade being released from the engine envelope. As in the blade-off tests above, the most adverse containment condition is calculated and demonstrated for fan blades by test in an engine. Similarly the capability demonstrated in the test is then 'read' into other engine conditions using modelling. For smaller compressor blades and turbine blades the containment systems tend to be simpler, just a thin metal casing as opposed to the combinations of metal and kevlar used for fan blades. With a simple metal thin skin containment the use of previous tests and experience reduces the amount of testing considerably.

6 SUMMARY AND CONCLUSIONS

The basis of the structural integrity process in the aerospace industry is as follows:

- A failure 'modes and effect' analysis is used to classify the failure modes.
- The integrity of non welded critical parts (defined as those whose primary failure could produce hazardous engine effects (e.g. uncontained debris, engine fires) at a rate in excess of extremely remote) is addressed by manufacturing process control and inspections.
- Probabilistic targets are employed to accommodate deviations between the inspection capabilities and the required component quality with regard to anomalies occurring in base materials.
- Certain components, e.g. fan blades, are designed to accommodate damage in service to a level specified by the airworthiness authorities but this does not guarantee that failure is prevented. The engine is therefore also designed to contain the release of a single blade.
- Certain critical components are designed for a life limit, e.g. discs, and are replaced when this limit is reached. Other components have some crack tolerance, and life extension can be achieved by regular inspection and repair of cracks and are therefore 'on condition'.

REFERENCES

1. Joint Aviation Requirements, JAR–E, 'Rules for the Certification of Engines', Change 10, Joint Aviation Authorities, Hoofddorp, August 1999.
2. Code of federal regulation, Aeronautics and Space, 14, Part 33, 'Airworthiness Standards: Aircraft Engines', US Government Printing Office, Washington, 2000.
3. Joint Aviation Requirements Part 25 'Large Aeroplanes', Change 15, Joint Aviation Authorities, Hoofddorp, January 2000.
4. BSI PD 5500/2000 Specification for unfired fusion welded pressure vessels, British Standards Institution, London, 2000.

CHAPTER 8

Methods for the Assessment of the Structural Integrity of Components and Structures: Civil Structures

F.M. Burdekin

Centre for Civil and Structural Engineering, UMIST, PO Box 88, Manchester M60 1QD, UK

ABSTRACT

A review is given of the wide ranging nature of civil engineering structures and the way in which structural safety is achieved. The main methods of achieving structural integrity at the design stage are through the use of well established procedures incorporated into codes and standards, using basic principles of structural engineering, estimates of 'worst case' loading, and incorporating appropriate safety factors. Modern Codes are generally based on principles known as 'limit state design' and use partial safety factors to give a target notional remote probability of failure which is considered acceptable, taking into account the consequences of failure of the particular type of structure.

The codes cover design against failure by plastic collapse, shear failure and buckling as standard guidance. Avoidance of potential failure by fracture in steel structures is usually achieved by selection of materials based on Charpy test requirements at the minimum temperature combined with requirements for quality of fabrication. Specific design guidance exists for avoidance of failure by fatigue in welded construction.

Where assessment of structural integrity is required to take account of the effects of unusual designs or loadings, effects of flaws or other circumstances with respect to potential failure by fracture or fatigue, the methods of BS7910:1999 are usually adopted. Examples are given of a number of cases where such assessments have been used to demonstrate acceptability or decide on remedial action for service of particular structures or components.

INTRODUCTION

Civil engineering structures are by their nature designed for a specific purpose usually as part of the infrastructure required by other industries or activities. Thus they are usually 'one-off' designs and may cost substantial amounts of

141

money and take lengthy periods of time to complete. Sometimes, but not always, the financing is provided from government sources, i.e. the public purse, and often the completed structure has to perform satisfactorily for many years without hazard to public safety. As a result the general principles of design are encapsulated in codes and standards for different structural applications.

Examples of different types of civil engineering structure include buildings, bridges, dams, tunnels, roads, railways, ports and harbours, offshore structures, power stations and leisure rides. For present purposes discussion will be restricted to buildings, bridges and leisure rides.

LOADINGS AND DESIGN PRINCIPLES

Civil engineering structures are potentially subject to a number of types of loading and the definition of these is essential in carrying out any structural integrity assessment. The loadings can be broadly divided into permanent effects (e.g. self weight), imposed loads (e.g. office equipment and people in buildings, machinery in factories, traffic on bridges), environmental loads (e.g. wind, snow, waves) and extreme loads (e.g. earthquake, typhoon). It may be necessary to give particular attention to situations where dynamic loads occur. Since the designer cannot control the future use of a structure, general recommendations are made for imposed, environmental and extreme loads in design codes based on the general nature of the structure and the location. Environmental and extreme loads are based on statistical records of the occurrence and magnitude of the events concerned at the location where the structure is to be built. For example, wind loading is based on the probability of a particular wind speed being exceeded within a time period of say 100 years, taking account of location, height above sea level, ground topography and interactions with adjacent buildings. Similarly the severity of earthquake loading is based on historical records of the probability of events of a particular magnitude occurring within a given time period, expressed in terms of maximum anticipated ground accelerations.

Two basic forms of design code exist, namely allowable stress and limit state design methods. For allowable stress methods, upper bound estimates of the loading are used to calculate stresses throughout the structure and members are proportioned so that the stresses do not exceed the yield (or occasionally ultimate) strength divided by an overall global safety factor. For limit state design methods, characteristic values of the loads are multiplied up by individual partial safety factors and the structure is designed to fail under these enhanced loads using material with a characteristic yield strength divided by a partial safety factor. The magnitudes of the partial safety factors are calibrated to give a target notional probability of failure allowing for the uncertainties in the loading and scatter in the material properties. The major advantages of limit state design are that it specifically considers different modes of failure allowing the use of different partial safety factors for each mode, it permits the effects of different levels of uncertainty to be allowed for, and it allows advantage to be taken of the effects of structural redundancy. Finite element analyses are extensively used to determine stresses throughout structures of any size or complexity.

MATERIALS

The traditional materials of the construction industry for structural applications are steel, concrete, masonry and timber. Many older civil engineering structures still exist in which either cast or wrought iron was used extensively and assessment of their structural integrity is particularly important when such structures are refurbished or converted for modern requirements. Use of composites is of increasing interest for civil engineering structures bringing with it different requirements for assessing structural integrity compared to the traditional materials. Plastics are now widely used for services and fittings such as pipes, gutters, window frames etc. Cladding for major buildings can represent of the order of one third of the total cost of the building and a very wide variety of materials is used for this purpose. This includes metals such as stainless steels, aluminium, bronzes, titanium and copper or non-metals such as glass, stone/masonry, cement/concrete panels and composites such as glass or carbon fibre reinforced plastics. Glass is now sometimes used structurally and is very extensively used for cladding of major buildings in many attractive ways with different colours and tints giving a delightful appearance. The different materials have their different mechanisms of failure so that assessment of structural integrity involves a full understanding of materials behaviour. A discussion of aspects of materials requirements for civil construction is given in the ICE Brunel 2000 Lecture.[1]

MODES OF FAILURE

The traditional procedures incorporated in codes for design of civil engineering structures are aimed primarily at avoiding failure by plastic collapse and buckling. Where structures are specifically subject to fatigue loading, as for bridges and cranes, design guidance is given based on the classification of different types of welded details and their experimental fatigue behaviour expressed in the form of S–N design curves. Avoidance of failure by fracture, where treated, is essentially covered by requiring selection of steels to have a minimum Charpy energy absorption at the minimum service temperature dependent on the thickness and yield strength of the steel. This is expressed alternatively by giving maximum thickness limits for different grades of steel (in terms of yield strength and Charpy properties) for different minimum temperatures. These requirements have been derived using fracture mechanics based structural integrity methods in some countries.

For example the requirements for selection of materials to avoid brittle fracture in the UK Steel Bridge[2] and Building[3] Design Codes have recently adopted the following formula:

$$t_{max} = k.50.\left(\frac{355}{f_y}\right)^{1.4}.(1.2)\left(\frac{T_{min} - T_{27}}{10}\right), \qquad (t_{max} \text{ in mm})$$

where t_{max} is the maximum permitted thickness in mm, f_y is the yield strength in MPa, T_{min} is the minimum service temperature and T_{27} the temperature for 27 joules energy absorption in the Charpy test (both in degrees Celsius), and k is a non dimensional factor to account for effects of stress level or stress concentration. For normal as-welded construction without stress concentrations, k is taken as 1.0, for low stresses or stress relieved structures k is 2.0, and for details involving for example cover plates or attachments welded to the edge of a plate the value of k is 0.7 or 0.5.

This formula was derived as an empirical fit to results of fracture mechanics analyses using the methods of BS 7910, with an assumed initial flaw depth of 0.15 times the thickness and using the Wallin correlation between fracture toughness and Charpy test values.

Design against fatigue failure in welded construction is normally based upon design curves derived from results of experimental tests on different geometries of weld detail. Guidance of this kind is given, for example, in the UK Steel Bridge Code[2] and in the specific code for design against fatigue in welded construction, BS7608.[4] For fracture requirements there is an inherent assumption that the quality of fabrication meets normal acceptance standards for weld defects although the codes do not include specific guidance on defect acceptance levels. Such guidance is included in some codes but it is usually based upon good work-manship standards rather than on a fitness-for-purpose basis. For fatigue design it is important to recognise that the normal design rules are based on a lower bound (mean -2 standard deviations) for results obtained on experimental tests of laboratory scale welded samples without defects.

Avoidance of failure by corrosion in civil structures relies on provision of appropriate corrosion protection although a significant number of corrosion problems still occur. There are also some potential problems with deterioration of concrete and masonry, such as sulphate attack, chloride effects, carbonisation, alkali–aggregate reactions, thaumasite reactions and freeze thaw problems.

STRUCTURAL INTEGRITY ASSESSMENT OF CIVIL STRUCTURES

ASSESSMENT OF STRUCTURAL INTEGRITY AT FABRICATION/CONSTRUCTION STAGE

Non destructive testing is commonly required on welded steel construction immediately after fabrication to confirm freedom from significant defects. Most defect acceptance standards are based on quality control requirements rather than fitness for purpose. Attempts have been made to produce such standards on a fitness for purpose basis but these have not been widely implemented. Structural integrity assessments are sometimes carried out at the fabrication/construction stage of specific welded construction to determine whether defects which have been located by NDT can be accepted, thus avoiding unnecessary repairs and delays. In these cases the most common approach is through the use of BS7910/PD 6493.[5]

ASSESSMENT OF STRUCTURAL INTEGRITY IN SERVICE

Buildings may be designed for anything from a few tens of years to several hundred years. A typical building life might be fifty to one hundred years although refurbishment will frequently take place at least once in the lifetime. Maintenance engineers will normally be concerned with building services rather than the structure of the building. The general structural condition of buildings is the business of Chartered Surveyors who are experienced in checking for symptoms of deterioration or damage. Occasionally specialist structural integrity assessments are necessary after extreme loading incidents such as earthquakes or explosions.

Bridges in the UK are designed for a life of 120 years. All highway bridges are required to have a principal inspection every six years. This is essentially a visual inspection with access available enabling all parts of the structure to be touched. Major bridges often have a permanent maintenance team and access provided through permanent gantries. In other cases access has to be provided by scaffolding, scissor or hydraulic hoists. Inspections of this kind show the condition of protective treatment, potential problems due to water ingress, and major faults sufficient to cause excessive deflection. Fatigue cracks may be found if staining of the protective treatment has occurred due to corrosion at the crack. Non destructive testing is rarely applied routinely in service, but is used if visual inspection has shown problems requiring further investigation. The general procedure for assessing structural integrity is to determine the current condition of the bridge and check against the requirements of a specific assessment code based on fitness for purpose with specialist assessments required only if significant defects are found.

Leisure/fairground rides are subject to stringent checks on structural integrity. Every ride has to be inspected visually before the start of operation each day. Major rides are required to have a full inspection annually by a competent independent engineer. In some cases the Engineer will require non destructive testing of critical regions.

The railway and aircraft industries are particular example where fatigue is of major concern to structures and components and public safety is of paramount concern. The basic principles of design against fatigue for railway applications are the same as for other applications except that for some components the number of cycles applied is so great that it is essential to operate below the fatigue limit. The dynamic effects from loading are such that there are uncertainties about stress levels and periodic inspections are necessary which are carried out by combinations of ultrasonic and magnetic particle testing. In some cases these inspection intervals have been decided on the basis of fracture mechanics assessments used to predict rates of fatigue crack growth.

BS7910:1999 ASSESSMENT METHODS

BS7910:1999[5] is the successor to the BSI Guidance Document PD6493:1991 originally developed in the 1970s and first published in 1980. This document is aimed at providing guidance on methods to determine the significance of flaws in metallic construction on a fitness-for-purpose basis. In its latest form, BS7910 includes the following sections:

- Types of flaw
- Modes of failure and material damage mechanisms
- Information required for assessment
- Assessment for fracture resistance
- Assessment for fatigue
- Assessment for plant operating at high temperatures
- Assessment for other modes of failure

For fracture assessments BS7910 has three alternative methods. At the first level, a simplified method is used based on the CTOD design curve and flow strength based collapse retaining continuity with the first version of PD 6493 in 1980.

The normal assessment method now recommended uses the failure assessment diagrams (FAD) of the R6 approach[6] with the alternatives of using either a generalised failure diagram (level 2A) or a material specific failure assessment diagram (level 2B). As with the standard R6 approach the user has to calculate the fracture ratio parameter K_r which is the ratio of applied stress intensity factor to fracture toughness and the plastic collapse parameter L_r which is the ratio of applied load to yield collapse load. When the points are plotted on the failure assessment diagram those lying inside the curve are deemed to be safe and those outside the curve are unsafe.

The third level for fracture assessment in BS7910 deals with ductile materials and allows a tearing analysis to be carried out. This level also provides three alternatives:

- 3A Generalised FAD
- 3B Material specific FAD
- 3C J-based geometry and material specific FAD.

A comprehensive series of Appendices is provided in BS7910 dealing with such issues as the following:

- Tubular joints
- Pressure vessels/pipelines
- Misalignment
- Leak before break
- Corrosion damage
- Weld strength mismatch
- Use of Charpy tests
- Reliability and partial safety factors
- Stress intensity factor solutions
- Residual stress distributions
- Proof testing and warm prestressing

BS7910 fatigue assessment methods are based on use of the Paris fracture mechanics law for growth of cracks from initial sizes to a final size. The Paris Law is integrated assuming standard values of the constants so that guidance can be given for fatigue life in the form of curves for different stress ranges for growth from any initial crack sizes to any final crack sizes in different weld geometries.

EXAMPLES OF STRUCTURAL INTEGRITY ASSESSMENTS

ASSESSMENT OF THE SPINDLE OF THE LONDON EYE

The London Eye is a major observation wheel constructed above the River Thames at Westminster in London (Fig. 1). The wheel is 135 m diameter with 32 observation capsules and takes approximately 30 minutes for each revolution. The wheel is supported at a central spindle which is itself kept in position from one side only by an arrangement of inclined columns and cable says. The spindle is thus a safety-critical element for the whole structure. The spindle is approximately 21 m in length and some 2 m in diameter with thicknesses of 200 to 300 mm. It was manufactured in Czechoslovakia from a series of cylindrical steel castings joined

Fig. 1   The London Eye with spindle at centre of wheel.

together with circumferential butt welds throughout the full thickness of the castings. Figure 2 shows the spindle after assembly and welding on delivery to site.

In view of the safety critical nature of this member a structural integrity assessment was carried out with particular reference to the castings and the regions of the circumferential butt welds. Fracture mechanics toughness tests were carried out on samples from the castings and welds and although the level of CTOD fracture toughness was generally satisfactory, occasional low results occurred. The stressing conditions for the spindle are relatively well defined and it should be noted that the spindle itself remains fixed with the wheel hub rotating on the spindle so that fatigue loading is not a major factor. Fracture mechanics calculations using the procedures of BS7910 suggested that for the occasional low toughness results relatively small defect sizes (less than 10mm) could be of concern. As a result of this two courses of action were decided. Firstly it was decided to focus non-destructive testing by magnetic particle and ultrasonic methods on the more highly stressed regions associated with the circumferential butt welds with an acceptance requirement of no detectable flaws. Secondly it was decided that the spindle would be subjected to a proof test to reduce residual stresses and give protection by loading it to a level of 1.2× the maximum design

Fig. 2 The spindle for the London Eye delivered to site.

loading to the same pattern as the service loading. This proof load was completed in a dockyard in Rotterdam using prestressing jacks and cables to achieve the very high loads required.

On successful completion of these measures the spindle was taken to site and completion of construction of the wheel and final erection took place.

ASSESSMENT OF MATERIAL AND WELD QUALITY FOR A MAJOR STEEL PLATE TRANSFER GIRDER

In the construction of a major building in Hong Kong the design involved a 75 m long by 5 m deep steel plate girder supported on three columns to provide clear spans over the ground floor foyer area (Fig. 3). These girders acted as the supports for the main columns of the remainder of the building at normal spacing intervals above this level. The design of the girder involved regions with both webs and flanges of 200 mm thickness in Grade 355 steel. In the region of the central column, a cruciform arrangement of plates occurred such that the column itself continued vertically up to the top flange of the plate girder, whilst the web and bottom flanges of the plate girder were split, with infill plates either side of the column (see Fig. 4). The designer had specified partial penetration butt welds within the lengths of the webs and flanges of the plate girder and partial penetration T-butt welds for the web to flange welds of different sorts. The original requirement was for a 25 mm depth of penetration in T-butt welds between the girder web plates and the column flange plates as shown in Fig. 5(a). Concern was expressed by the independent design check authority as to whether partial penetration butt welds of this kind would be satisfactory.

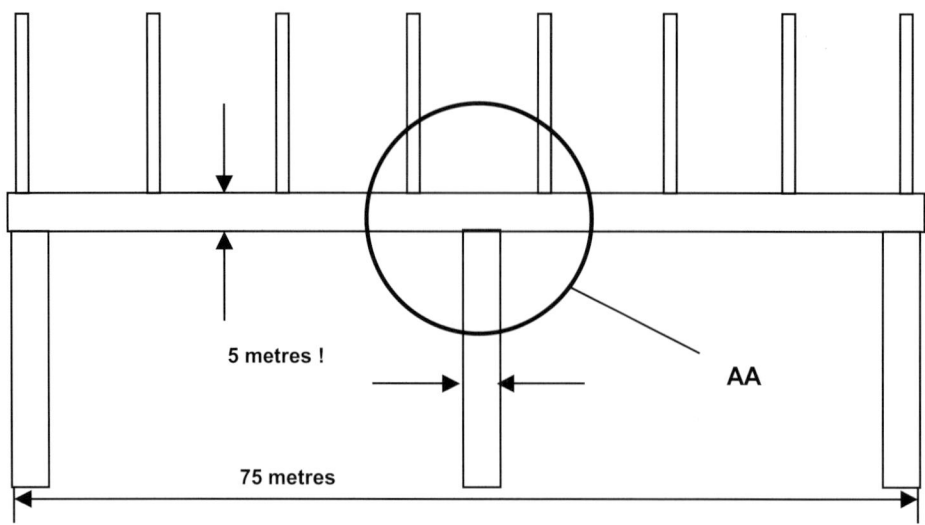

Fig. 3   Overall arrangement of transfer girder.

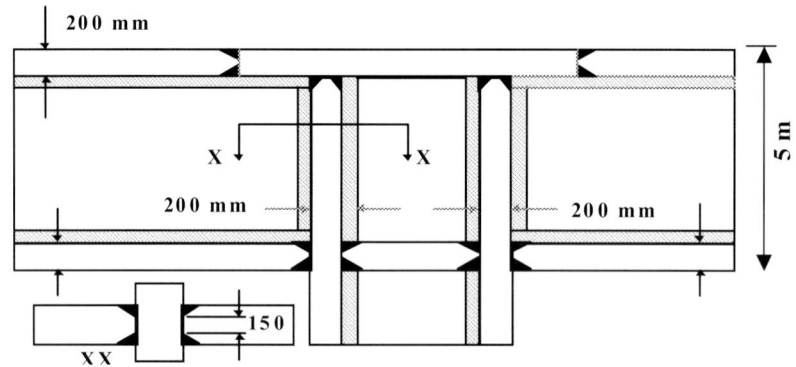

Fig. 4   Transfer girder over central column (section AA).

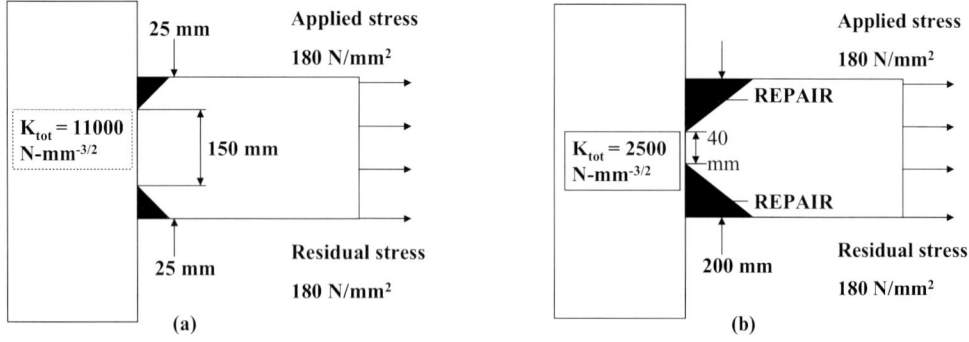

Fig. 5   (a) Web/Flange detail as designed; (b) modified web/flange detail.

An independent structural integrity assessment was carried out treating the unpenetrated region of the web to flange welds as a crack-like defect rather than just considering the throat area of the welds. Assumptions had to be made about residual stress levels and for this type of connection it was assumed that they would be about half yield strength level. The applied tensile stresses across the top of the central column section were also about half yield. Calculations of the applied stress intensity factor for a defect of 150 mm height in a thickness of 200 mm show K values of about $11000 \, \mathrm{N \, mm}^{-3/2}$ ($350 \, \mathrm{MPa}\sqrt{\mathrm{m}}$). This was considered well above any credible fracture toughness levels for the weld metal for the submerged arc process used to manufacture the web to flange welds. It was therefore decided that partial penetration welds of depth 25 mm in regions of tension stress of half yield presented an unacceptable risk of fracture and repairs would have to be carried out. The repair procedure specified was to excavate the existing partial penetration welds and re-prepare them to a depth of 80 mm from each site, reducing the unpenetrated flaw height from 150 mm to 40 mm (see Fig. 5(b)). Fracture mechanics calculations for this flaw size with the same stressing conditions showed that the applied stress intensity factor had been reduced to $2500 \, \mathrm{N \, mm}^{-3/2}$ ($80 \, \mathrm{MPa}\sqrt{\mathrm{m}}$) and it was considered that the fracture toughness of the weld metal would be significantly greater than this. This repair method was applied to all partial penetration butt welds in tension regions to ensure that the remaining unpenetrated region was acceptable.

It should be noted that this represents an example of the use of as welded construction for plate thicknesses well in excess of those which would be permitted in any normal material selection code requirements. The Charpy test properties of the plate material were known to be excellent, with approximately 100 J at $-40°\mathrm{C}$. The fracture mechanics assumptions demonstrated that adequate fracture toughness should be available in the weld metal and parent material for the stress levels and flaw sizes present including allowance for residual stresses. Whilst use of partial penetration welds in tension requires significant caution, this example demonstrates that an economic solution can be obtained by modern structural assessment procedures.

ASSESSMENT OF DEFECTS IN AN AIRPORT ROOF STRUCTURE

In the construction of a major international airport the roof of the main terminal building was designed to be a space frame made up of tubular steel members. These tubular members were of the order of 1 m diameter and 40 mm thickness and were fabricated from plate material with longitudinal seam welds, and it was found that transverse crack sizes up to 15 mm width $\times$ 15 mm depth had occurred. The only fracture toughness data available suggested Charpy test values of 40 J at $0°\mathrm{C}$ for the weld metal. The yield strengths of the pipe material and weld metal were 350 MPa and 400 MPa with applied stresses of 250 MPa in the most highly stressed areas.

A structural integrity assessment to BS7910 showed that for surface defects 15 mm × 15 mm, values of 1.13 for K_r and 0.63 for L_r would be obtained indicating that the situation was not acceptable. The critical flaw sizes calculated using these methods were about 9 mm × 9 mm.

The client was advised that the best line of action would be to proof load the roof structure but he was reluctant to do this because this would demonstrate a lack of confidence in the roof structure to any observers. The next approach was to attempt a critical crack arrest assessment to see whether the pipe material would arrest any cracks initiating in the weld. The Charpy properties of the pipe material were reported as 100 J at 0°C, corresponding to a dynamic fracture toughness value of about 140 MPa$\sqrt{m}$. Fracture mechanics assessment for the size of crack equal to the width of the weld metal (approximately 30 mm) showed results which were again not acceptable. To assess the conservatisms in these procedures it was decided to carry out a series of crack arrest tests at the minimum service temperature.

Lengths of tubular member were prepared in which the existing longitudinal weld was replaced by a brittle weld deposit over a length at the centre of the specimen, as shown in Fig. 6. Separate crack initiators were welded on top of the brittle deposit and the region cooled locally to force initiation of fracture in the brittle deposit. With pre-existing cracks in this brittle weld, the tubes were supported in three point bending and loaded dynamically with the cracked brittle weld in tension. In all cases initiation of fracture in the welds occurred and the cracks arrested in the parent tube material.

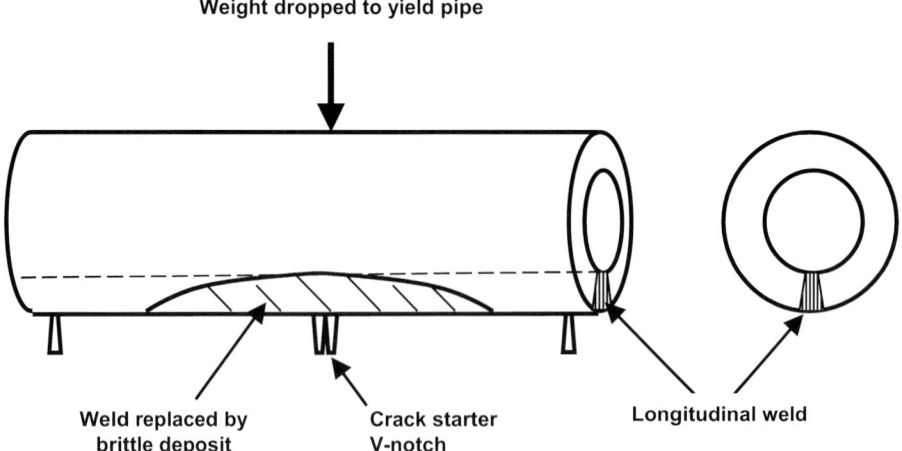

Fig. 6 Crack arrest test on tubular member for airport roof structure.

Having demonstrated that a crack arrest argument could be supported experimentally if the worst combination of crack size, toughness and stress levels allowed initiation, the client accepted the structure as suitable for service.

CONCLUSIONS

1. Civil Engineering structures are individual, diverse and very expensive.
2. Standard design procedures achieve required safety levels through the use of partial safety factors.
3. Non-destructive testing is applied for quality control after fabrication.
4. Fracture mechanics based procedures such as BS7910:1999 are used for special cases either at the design stage or when defects are discovered in service.
5. Examples have been given of the application of fitness-for-purpose assessments for three different types of civil engineering structure for which the solutions were proof loading, repair and crack arrest testing in addition to standard assessment methods.

REFERENCES

1. F. M. Burdekin, Institution of Civil Engineers, Brunel International Lecture 2000, ICE, July 2000.
2. British Standard BS 5400, Pt. 3, Design of steel bridges, British Standards Inst., 2001.
3. British Standard BS 5950, Design of steel buildings, British Standards Inst., 2001.
4. British Standard BS 7608, Fatigue of welded structures, British Standards Inst., 1993.
5. British Standard BS 7910, Guidance on methods for accessing the acceptability of flaws in fusion welded structures, BSI 1999.
6. I. Milne, R. A. Ainsworth, A. R. Dowling, and A. T. Stewart, Assessment of the integrity of the structures containing defects, British Energy R6 Revision 4, 2001.

CHAPTER 9

Applications of Structural Integrity Assessments for the Offshore Industry

C. S. Wiesner and G. R. Razmjoo
TWI, Granta Park, Abington, Cambridge, CB1 6AL, UK

1. INTRODUCTION

Mechanical failures in the offshore sector of the oil and gas industry are fortunately rare. However, if they occur, their ecological and economic consequences can be most severe. With increasing understanding of the mechanical behaviour of structures and improving material properties, this industry sector is therefore striving for the continued reduction in failure rates through the application of better predictive material and process models. At the same time, increased competitiveness, cost awareness and the ageing of infrastructures have led the industry to invest in R & D developments to justify reduced safety margins, more cost-effective design methods and extensions to design life.

Relevant structural integrity work includes applied research into the material property, performance and inspection of offshore structures and components, based on technical capabilities and experience relating to metallurgical behaviour, fracture, fatigue, design, residual stresses, failure investigations, advanced NDT, plant assessment and materials testing often referred to as fitness-for-service or engineering critical assessment (ECA).

Engineering critical assessments which have been standardised (e.g. Ref. 1) through much R & D and committee work in the last three decades or so can be successfully used to support the above goals throughout the life cycle of an offshore structure:

- In the concept and design phase, material property and dimensional requirements can be set and innovative design approaches can be justified.
- During fabrication, the significance of welding or manufacturing flaws can be evaluated and the need for repair can be assessed.
- During routine in-service inspection, ECAs can be used to decide whether or not repair is required, or whether service conditions can be investigated so that repair can be delayed until an opportune moment or forever.

155

- Towards the end of the design life, assessments of the structure's state can be carried out using ECA principles to justify life extension and re-qualification.

By focusing on the factors which, if neglected, could increase failure probabilities the competent application of ECA will result in safer structures with fewer repairs, better-focused NDT and the possibility of relaxed acceptance criteria (compared to workmanship standards).

The present paper summarises examples of the work on offshore structures such as fixed jackets; mobile drilling units; pipelines; risers; and floating production, storage and offloading (FPSO) structures.

2. ENGINEERING CRITICAL ASSESSMENTS FOR FIXED JACKETS, SEMI-SUBMERSIBLE RIGS AND PIPELINES

2.1. MAJOR OFFSHORE JACKET STRUCTURE

This is an example where engineering critical assessments were used to demonstrate the acceptability of flaws whose dimensions were outside the acceptance criteria as specified in the relevant fabrication code.

At a late stage in fabricating a major offshore platform, extensive chevron cracking (weld metal hydrogen cracking) was found in submerged-arc longitudinal seam welds in the tubulars used to make the nodes. There was a crack about every 50 mm along the welds. The tubulars, which were up to 100 mm thick had been made by one subcontractor. A second subcontractor had welded brace stubs to these to make the nodes, which were then post-weld heat treated (PWHT). The main contractor had received the nodes and welded them into the jacket. The chevron cracking was first found during the ultrasonic inspection (UT) of one of the circumferential erection welds, when the NDT technician had to probe through the end of the longitudinal seam of the node, where this intersected the erection weld. Further UT probing revealed that up to 20% of seam welds were affected.

Repair of the cracks would have had severe consequences for the construction programme. The structure would have had to be dismantled, because *in-situ* PWHT of the repair welds was impossible. As for the Forties field nodes, a fatigue and fracture engineering critical analysis showed the flaws to be quite harmless. Based on the analysis, it was decided, with the concurrence of the certifying authority to allow construction to proceed. This decision is believed to have saved the company at least £45M. A number of lessons were learnt from this experience:

- Naturally, it would have been better if the flaws had been avoided initially, but it was not known that they existed until very late in the day and they were not found on two earlier inspections.

- In the light of this and of many other experiences, it is unrealistic to rely (for structural safety) on achieving 'clean' weld metal.
- CTOD pre-qualification of the weld metal was invaluable. Without this, there would have been a one month delay for the tests to be performed, during which period, construction would have been halted.
- The cost of achieving good toughness in these node welds, the critical welds in the structure, was negligible compared with the possible consequences of the weld metal being brittle. This is put into context by Cotton[2] who estimates the total cost of node fabrication at only 0.75% of the installed platform cost.

Other examples of this type of ECA application are given in Refs 3 to 9.

2.2. LIFE EXTENSION OF SEMI-SUBMERSIBLE RIGS

This example shows the use of ECA to demonstrate structural integrity for continued operation.

A number of semi-submersible drilling rigs are reaching the end of their original design lives. The design and fabrication costs for such rigs has multiplied many-fold since the 1960s. Extensive analysis and refurbishment programmes including ECAs to provide additional design life have been carried out for many rigs at approximately 10% of the price of a new rig. The structural integrity part of these programmes[10–13] involved the following (see Fig. 1).

- Fracture toughness tests and ECAs to predict the tolerable final flaw size.
- Overall stress analysis of the rig and detailed stress analysis of the joints to identify regions of high fatigue loading.
- Inspection programmes targeted on these regions.
- Remedial measures (burr grinding) of suspect regions and those subjected to severe fatigue loading.

Examples of life extension work for other welded structures and components are given in Refs 4 and 14 to 17.

2.3. OFFSHORE PIPELAY

This case is an example of the optimum way in which ECA can be used during the design stage of a structure. An engineering critical analyses for this 50 km line was carried out 'up front', before laying commenced in order to develop flaw acceptance criteria based on fitness-for-service principles rather than workman-ship limits. There were two outcomes from this analysis.

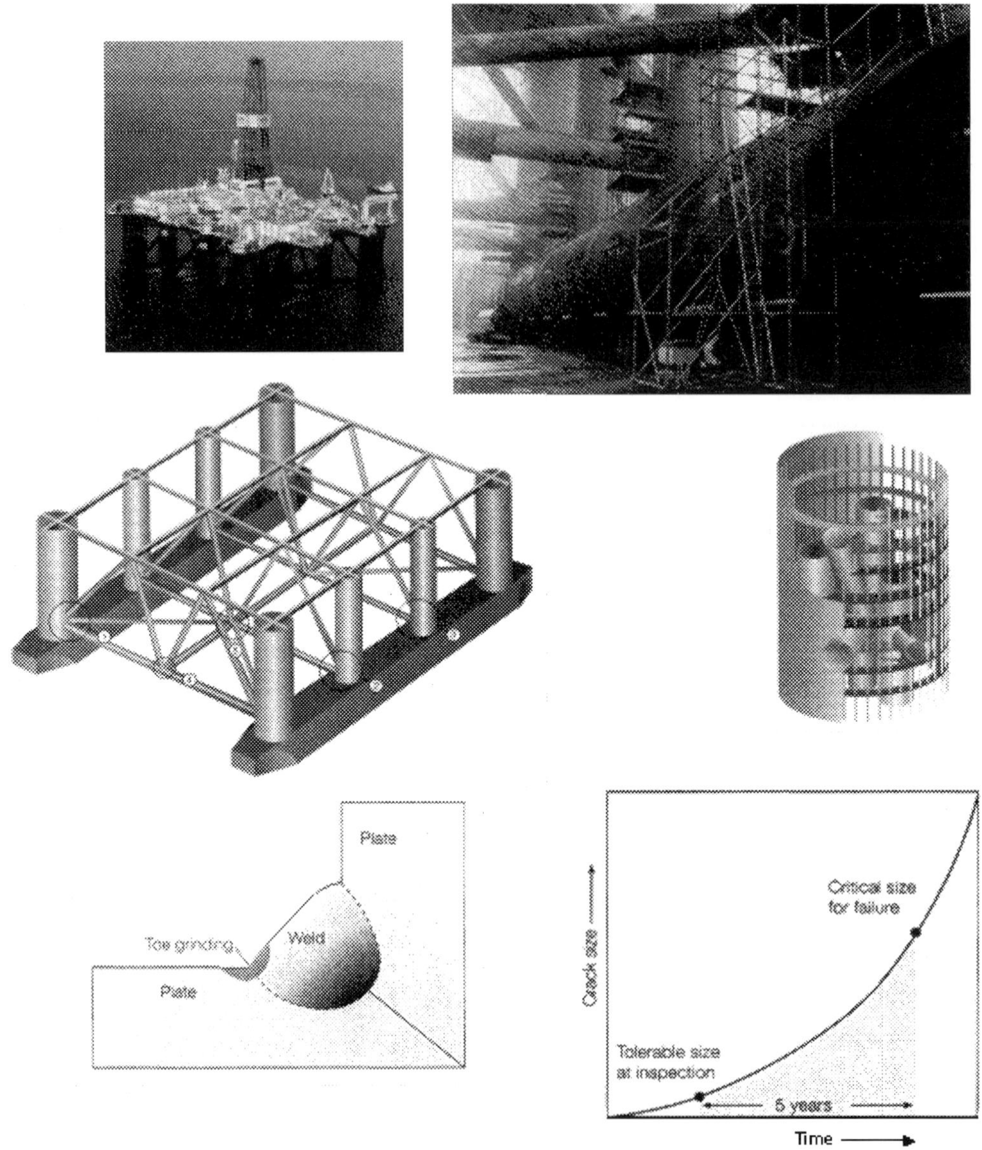

Fig. 1   Semi-submersible layout, analysis and fatigue life improvement principle.

- It was decided to use mechanised UT on the barge, rather than radiography. UT is faster and provides more reliable flaw depth dimensions.
- ECA-based flaw acceptance criteria were established in place of the conventional workmanship standard.

These two decisions resulted in record pipelay rates exceeding 4 km/day. They saved about 11 days barge time, at about £250 000 per day. The flaw acceptance criteria reduced the repair rate to 5% of that which would have been required had workmanship acceptance criteria been imposed. This is despite the fact that the UT detected more imperfections than would have been found by radiography. The total saving was at least £3M.[18]

It is also worth noting that the analysis approach has been used to justify the adoption of mechanised MIG welding for offshore pipelining. This method might otherwise be rejected because of its tendency to introduce 'cold shuts' or lack of sidewall fusion. Mechanised MIG welding can save as much as $60 000/km of offshore line, compared with conventional manual stove-pipe welding.[4]

Other examples of structural integrity assessments during the material selection, design stage and for repair are given in Refs 4, 19 and 20.

3. STRUCTURAL INTEGRITY WORK FOR DEEPWATER EXPLORATION AND MARGINAL FIELDS

3.1. OPENING REMARKS

The depleting reserves in shallower water and greater potential for larger finds elsewhere, more favourable fiscal policy, and impressive new technology, are the key factors driving operators to produce oil and gas from marginal fields and at greater depths. It is expected that most new fields will be discovered in waters deeper than 200 m. Although deepwater exploration and production has been occurring in several locations worldwide for many years, the current use of floating production systems and new materials offshore have created significant challenges in their design, welding fabrication and inspection. This section describes the structural integrity related issues and highlights some of the current research.

3.2. INTEGRITY OF FLOATING PRODUCTION, STORAGE AND OFFLOADING (FPSO) VESSEL HULLS

Ship-based floating production, storage and offloading (FPSO) vessels are a relatively new concept, with great potential for exploring marginal and remote fields. These vessels may be purpose-built or converted tankers, depending on the economics of the field which they service. It is expected that in-service damage (e.g. fatigue cracking) to FPSOs is more critical than to conventional tankers. The structural integrity of their welded joints must therefore be fully evaluated and understood to avoid any in-service failures and costly repairs.

A ship-based FPSO unit includes the vesssel's hull (storage); deck processing equipment; the turret, swivel and mooring system; and manifolds, flowlines and a subsea infrastructure (wellheads, Xmas trees, etc). There are economic advantages in using FPSOs for development projects; and both new-build hulls and converted tankers are considered when developing the business strategy (capital cost and time to first oil) for such fields.

The frequency and type of in-service damage in tankers shows a strong dependency on age because such port-to-port conventional operation tankers spend a considerable proportion of their lives in calm water and can avoid poor weather. They are also subjected to regular structural surveys in dry dock. FPSOs on the other hand are semi-permanently moored and may experience rough sea conditions for much of their lives. This suggests that welded hull details in FPSOs are more severely loaded than those in conventional tankers, and there are therefore important differences in the structural integrity requirements between an FPSO and a tanker. Ship design has historically been partly empirical, since ship failures are rarely caused by one single factor, and the fatigue design methods for tankers have evolved in the light of experience. However, the concept of the turret-moored FPSO, weather-vaning on its turret, is relatively new and service history cannot provide a reliable guide to design.

Work was carried out to assist an operator to develop material property requirements, fatigue life improvement and upgraded inspection procedure during design and fabrication of a FPSO vessel for a recently-completed North Sea project.[21,22] In-service inspection procedures were also derived using fracture control and fatigue assessment programmes[23-25] based on BS7910 procedures[1] with the objective of extending the period without dry docking to 10–20 years. The fracture and fatigue control programme (FFCP) estimated maximum tolerable flaw dimensions and the expected fatigue crack growth of critical components. An experimental comparison of two fatigue life improvement techniques (weld toe grinding and hammer peening) showed that the most convenient technique in terms of cost and quality control can be selected as both gave comparable increases in fatigue properties.

The work undertaken has demonstrated that a FFCP has many advantages on the selection of the appropriate material, welding process and filler metals for the construction of a FPSO hull, which could also be applied for oil tankers and any other sea-going structure. It has also been demonstrated that the FFCP can be used very effectively in the design and construction stage of the vessel. The accuracy of the method depends on the quality of the input data available for the analysis. Therefore, the data determined in the earlier stages of design, construction and in-service inspection will improve its effectiveness throughout the service life of the vessel. The full benefits of a combined in-service inspection and FFCP to mitigate the risk of catastrophic failure and to improve the maintainability of the hull structure, in order to increase the operation period without the need for docking, will become apparent during the operation of the vessel.

3.3. INTEGRITY OF TLP TENDON AND RISER GIRTH WELDS

3.3.1. *Background*

The welded joints in floating production systems, such as girth welds in risers of deep draft caisson vessels and tendons in tension leg platforms (TLPs), are more critical than those used in conventional jacket structures. The restriction to single-sided welding, in conjunction with the less stringent dimensional control inherent in seamless pipe manufacture, make the welding and inspection for risers more difficult and less reliable. Risers in particular are subjected to severe fatigue loading and have no redundancy. The failure of one weld will result in failure of the concept. Even the development of a through-wall crack, let alone complete rupture, could have serious environmental, safety and cost implication. It is therefore crucial to establish the inspection reliability of the girth welds at the fabrication stage and determine the fatigue performance of typical girth welds.

3.3.2. *Tension Leg Platform Tendons*

Tension Leg Platforms (TLPs) are partly-submersible structures, deployed to service reserves in deep water. The hull, constructed in either steel or concrete, is tethered to the seabed by clusters of girth-welded steel tubular tendons at each corner of the structure (see Fig. 2). TLP tendons are made onshore; either all-welded and towed out, or fabricated in several joint lengths onshore and assembled offshore using mechanical connectors. TLP tendons are made from relatively large diameter welded pipe, in which girth welds can be produced and inspected from both the inside and outside.[26]

The TLP tendons are subjected to severe fatigue loading in service. Tendons store a great deal of energy, and failure would lead to major damage to the platform, template and possibly the wells. In comparison, failures of tubular members in conventional jacket structures may have less drastic consequences because of redundancy. The fatigue performance of tendons in the as-welded condition was evaluated by testing a number of full-scale girth welded specimens.[27] Since TLP tendons are made from pipes of relatively large diameter the weld caps on the

Fig. 2 Tension Leg Platform (TLP) secured by tendons to the seabed.

inside and outside are usually ground flush to enhance the fatigue performance and to facilitate inspection. The European fatigue design classification for these girth welds is Class C. However, this classification is not based on experimental data for girth welds and there is concern that it will not be readily achieved as a result of the increasing significance of small welding flaws when a high fatigue strength is demanded. Recent research[28] has provided additional data. Ongoing work on fatigue testing full scale girth welds representative of TLP tendons will provide additional information on this subject.

3.3.3. *Risers*

Risers deployed from floating structures are either of the flexible type or of welded pipes. The welded risers, which are generally installed on economic grounds, can be in the form of Top Tension Risers (TTRs) or Steel Catenary Risers (SCRs). TTRs are girth-welded tubulars fabricated onshore and joined together by mechanical connectors offshore. They are deployed vertically from a floating structure and are connected to hydraulic tensioners at the top end, and to a weld head (for production risers), or to a template (for export risers), at the seabed.

In contrast, SCRs are simply the continuation of pipelines that hang from the floating structure at the top end and rest on the seabed at the lower end (see Fig. 3). SCRs are girth-welded pipes, fabricated offshore and installed from a barge in a J-lay (near vertical) formation by welding the pre-fabricated joints in the 2G position (where the pipe is held vertically with welding heads, or welders, revolving around the pipe). The relative simplicity of SCRs and the fact that they avoid the need for templates at the seabed and hydraulic tensioners at the top have made their deployment an attractive proposition. However, like TLP tendons, SCRs are

Fig. 3. A spar production platform showing two steel catenary risers and a typical arrangement of mooring lines.

subjected to severe fatigue loading. In addition to the lateral movement resulting from the vessel and wave motions, they may be subjected to a large number of cyclic loads due to vortex-induced vibration caused by ocean currents. SCRs in floating structures are further subjected to increased heaving. This motion will produce severe cyclic loading at the touch down on the seabed where the riser is curved and is joined to the pipeline.

SCRs are typically produced with small diameter seamless pipes, welded from the outside. The European fatigue design codes for offshore structures penalise butt welds made from one side. The design guidelines for these welds (Class F_2) are more severe than for welds made from both sides (Class E); unbacked one-sided girth welds are downgraded by two design classes even if full penetration is assured. This represents a reduction in allowable stress of 25% compared with the potential for butt welds made from both sides. None of the existing rules are based on fatigue test data for girth welds and there has been a need to characterise the fatigue of single-sided welds by testing full-scale specimens. Work is ongoing in a 'joint' industry project initiated in 1996 to establish fatigue design classification for SCR welds. This work has shown that single-sided welds can have good fatigue performance, as long as significant weld root flaws are avoided (see, e.g., Ref. 29). However, the applicability of these findings ultimately depends on the reliability of the weld inspection technique which is further outlined below.

3.3.4. *Inspection reliability*

If good access can be gained to both the inside and outside of a girth weld, inspection can be reasonably straightforward and reliable. Standard surface inspection methods as well as volumetric non-destructive evaluation (NDE) methods are all feasible. For the case of single-sided welds with limited access to the inside, reliance is placed on volumetric NDE methods to inspect the weld root, and ultrasonic testing (UT) is generally used for this purpose.

To achieve the desired fatigue life in a critical girth-welded component like a riser, high-quality welds are produced which require a correspondingly high quality of inspection. Most operators assume that since very small flaws can be shown to be detectable by NDE, only very small flaws will remain after the inspection. Unfortunately, this is not the case. Answering the question: 'How small a flaw can be detected?' does not answer the question: 'How large a flaw can be missed?' For quality control NDE, the first question is pertinent. However, for fitness-for-purpose NDE, the question 'How large a flaw can be missed?' is key (particularly for critical non-redundant SCR joints where failure from one weld flaw could result in the failure of the entire system). Quantifying the first question is relatively straightforward, requiring a few laboratory trials to establish signal and noise level. Quantifying the second question is much more difficult, since it involves 'human factors', and controlled trials on statistically significant numbers of fully representative test pieces.

The detection and sizing of volumetric and planar flaws at an irregular weld root is not easy. The inspection method should be carefully designed to discriminate between the echos arising from the weld root as distinct from a flaw. For joints of low to moderate thickness, this can be achieved with angled pulse echo probes used in raster-scan fashion. If such a techniques can be automated, to produce computer images and permanent records of the weld inspected, discrimination between the root and flaws may be more reliable. However, unpublished work at TWI in the late 1990s showed that even with optimum inspection procedures, there may still be only a 50% chance of finding flaws which are 2 mm deep × 12 mm long when subject to ultrasonic testing from the outside. Such a weld root defect can significantly reduce the fatigue life of the joint, potentially from Class E to Class F_2; i.e. a reduction of 25% in the allowable stress range, see Fig. 4.

The actual reliability of the NDE technique will depend on many parameters such as wall thickness, fit-up tolerances, and weld process. As a consequence, shallow root flaws with depths in the order of <0.5 mm are undetectable by ultrasonic testing from the outside of the pipe. Of course, applying more than

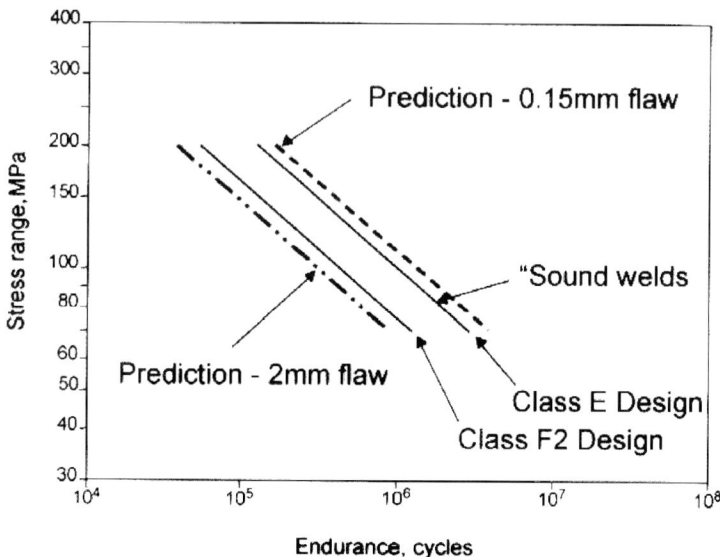

Fig. 4 Fracture mechanics-based life predictions for tubular structures for various initial flaw depths, superimposed on the S–N design curves for Class E and F_2 details.

one NDE method may reduce the size of flaws that might otherwise be missed. For example, the weld root could be subject to radiography. Indeed, for critical single-sided girth welds where a high degree of confidence in the detection of planar flaws in the root is required, the principle of 'diversity and redundancy' should be adopted: i.e. use as many NDE methods and techniques as possible. However, this will significantly and adversely affect the economics of making such joints.

Work currently in progress aims at optimising current automated ultrasonic testing (AUT) techniques for the inspection of the root of single-sided girth welds. The work will quantify the reliability of crack detection and sizing in the weld root. A library of girth welded specimens with weld flaws have been established to accomplish this.

4. PROBABILISTIC STRUCTURAL INTEGRITY EVALUATION RELEVANT TO THE OFFSHORE SECTOR

4.1. INTRODUCTORY COMMENTS

There is a clear trend in recent years towards the increasing application of statistics and probability in relation to the assessment of structural integrity. This trend is moving design, manufacturing and plant integrity management decisions away from lower bound deterministic approaches and towards decisions based on the balance of probabilities where uncertainties in the available data are treated statistically. In support of this trend, new statistical methods and applications of existing methods to offshore structural integrity problems are required. The following sections outline three examples of recent probabilistic and reliability structural integrity work.

4.2. PROBABILISTIC FATIGUE CRACK GROWTH ANALYSIS FOR CATENARY RISERS UNDER VARIABLE AMPLITUDE LOADING

In recent times, flaw acceptance criteria are increasingly derived using fracture mechanics analyses. In most cases, the analysis is deterministic, based on worst-case assumptions of every input parameter. This can result in overly conservative acceptance criteria and unnecessary weld repairs. Recent work has therefore focused on the development and application of probabilistic approaches.[30]

A probabilistic fatigue crack growth analysis provides a rational means of estimating the failure probability associated with particular flaw dimensions. This implies that acceptance criteria can be set in order to limit failure probability to a specified target value, depending on the degree of uncertainty of the variables. An improved probabilistic fatigue crack growth method[31] was recently developed for the assessment of structures subjected to variable amplitude loading. Existing probabilistic approaches have been enhanced with regard to the treatment of fatigue crack growth threshold where a structure is subject to variable amplitude loadings that do not follow a standard statistical distribution. Specific aspects of the method, including accuracy of crack growth estimate and crack shape changes under loading, are verified by comparing results with those obtained from deterministic fatigue analyses to BS7910.[1]

The model has been applied to steel catenary risers to determine the probability of an initial flaw growing to a specified final size as a function of time. Sensitivity studies were used to assess the effect of the crack growth threshold, initial and final flaw sizes, and uncertainty in the stress modelling on the probability–time trajectory. The method developed provides a means of determining the largest single flaw that may remain in the weld at installation without the failure probability in service exceeding a specified target value (see example calculations in Fig. 5). The results of the example calculations were used to generate plots of failure probability against time. Tolerable initial surface flaw sizes were calculated using the probabilistic method illustrating how the method could be used as the basis for weld flaw acceptance criteria.

The sensitivity of results to input assumptions including the final tolerable flaw height, stress modelling error and the effect of crack growth threshold were demonstrated. Results were found to be significantly sensitive to these variables;

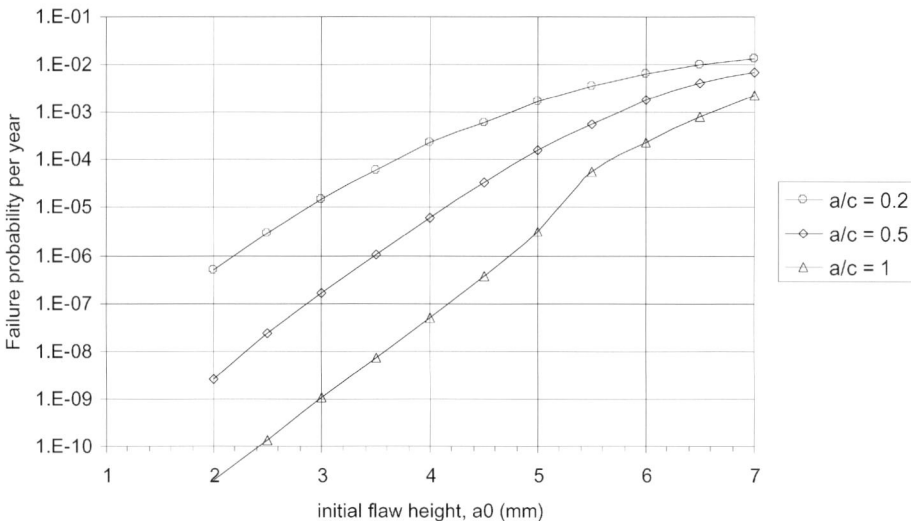

Fig. 5 Variation of annual failure probability with initial flaw height for a 100 year design life and an assumed final flaw height of 50% the wall thickness.

this highlights the need for appropriate data if practical conclusions are to be drawn from these analyses. The probabilistic approach provides a framework within which larger tolerable flaws may be safely justified.

4.3. EXTRAPOLATION OF THICKNESS MEASUREMENT DATA FOR NORTH SEA PIPELINE SYSTEM

The condition of a pipeline system on the North Sea oil platform has recently been assessed using an ultrasonic corrosion mapping technique. However, part of the system is inaccessible for inspection. In this region, therefore, it was necessary to develop a method to estimate the condition of the pipe based on sample inspections in the accessible area. This was achieved by fitting statistical distributions for the minimum wall thickness to the sample data. These 'extreme value' distributions were then used to derive theoretically the corresponding distributions in the inaccessible part of the system.[32]

Prior to inspection, the overall condition of the pipeline system was unknown, and corrosion attack was suspected. For the derivation of the thickness distributions it was assumed that: (i) the wall thickness in each uninspected region of the system followed the same distribution as some region that had been inspected; (ii) at the start of life, the mean wall thickness was midway between the manufacturing tolerances; and (iii) the ultrasonic measurement errors were negligible.

The most suitable fitting methods depends on the data, and the raw inspection results in this case did not fit any common distribution. If a satisfactory common fit to the underlying distribution cannot be identified, then it is usually possible to instead fit an extreme value distribution directly to minimum wall thickness measured over rectangular 'blocks' of a certain fixed size. This method was suitable for most of the inspection data.

Most of the statistical theory of extreme values is based on the assumption (sometimes implicit) that individual thickness measurement are statistically independent or, at least, that any correlation between the data is negligible. Stronger correlation between (for example) adjacent data points is described as 'clustering'. This clustering effect is mitigated, to a certain extent, by reducing the sample to that of the block minima. Each dimension of the block is chosen such that pairs of data points separated by the distance (in the appropriate direction) are weakly correlated. It can be judged how well a given extreme value distribution fits the data by examining whether the block minima show a linear trend when plotted on an extreme value probability plot. In this case, the data show a good fit to an extreme value distribution, see Fig. 6. Statistical software can then be used to estimate the location and scale parameters of the distribution, using the maximum likelihood method. Once the distribution is established, the methodology for extrapolating the distribution of minimum thicknesses over area uses

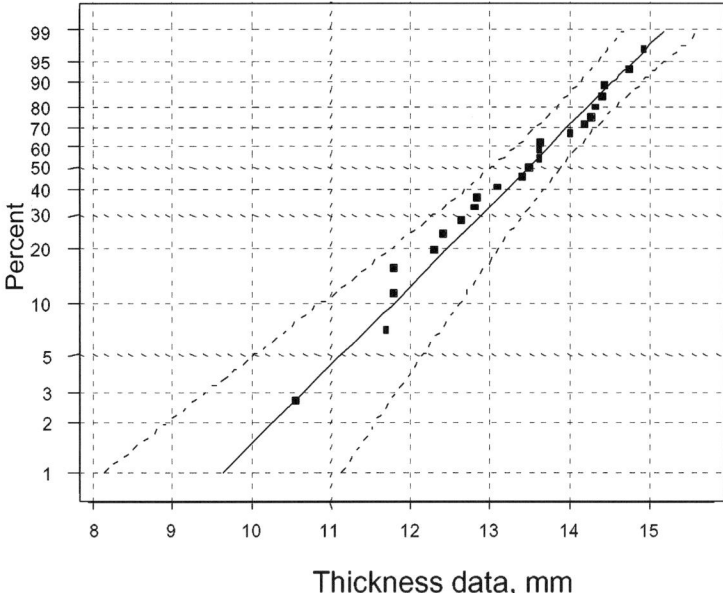

Fig. 6 Type I extreme value probability plot of minimum thicknesses of $0.03\,\text{m}^2$ blocks in the most corroded region of the tie-in pipe.

well-established techniques. When extrapolating to a pipe having a reduced initial thickness (relative to that inspected), the extrapolation factor used is reduced by the difference in the (mean) initial thicknesses. Extreme value statistics provide powerful tools for extrapolating sample inspection data into uninspected regions of a component. These methods have been developed to predict remnant life of pipelines in uninspectable areas. Other examples of recent probabilistic inspection and NDT research are given in Refs 33 to 36.

4.4. MODELLING OF HAZ HYDROGEN CRACKING PROBABILITY IN C–Mn STEELS

There are many approaches used worldwide to derive welding procedure for C–Mn steels to minimise the risk of HAZ hydrogen cracking during fabrication. They all have an allowance for 'safety' included, but this is not necessarily quantified. Determining the overall probability of cracking would allow control measures and inspection following fabrication to be concentrated in critical areas, thereby improving cost effectiveness, and would further provide an input into safety case assessments.

The TWI Nomogram[37] is one of the principal methods employed for welding procedure derivation. The nomogram limits were originally determined to give a 99.5% probability for 'safe' hardness levels (i.e. which would prevent cracking). To develop the probabilistic approach, the composition/cooling rate/hardness equations underlying the nomogram were identified and an analytical approach was developed. Comparison was made between predicted critical heat input levels to avoid cracking for various carbon equivalents and two thicknesses. The validity of the calculations was assessed first by comparing the calculated critical heat input levels with those obtained directly from the nomogram. A probabilistic model based on comparing distributions of hardness prediction against critical hardness levels for cracking was developed using commercially available reliability software. The model was then employed to determine whether or not the calculated heat input value corresponded to 0.5% probability of exceeding the critical hardness.[38]

Using controlled thermal severity (CTS) data generated from 80 separate test series in previous work, threshold conditions for HAZ hydrogen cracking were determined and employed to compare the traditional and probabilistic approaches. The threshold conditions were also utilised to determine the distribution of critical hardness associated with cracking for a particular hydrogen level. Using this distribution, the probability of cracking was determined for all test points in selected CTS series.

Application of reliability calculations to evaluate the range of conditions indicated that the method was working correctly. Given the experimentally determined distribution of critical hardness for cracking, the reliability calculations could successfully estimate the probability of cracking in the CTS test series examined.

The inputs to the reliability calculations which are considered as deterministic are composition, preheat, thickness and heat input. The last three generate a deterministic cooling rate value which, with composition, is used to predict a hardness value. Comparison of such calculations with experimental data has shown that the predicted hardness is the mean of a normal distribution with a standard deviation of 28 HV. The critical hardness for cracking to occur is a function of hydrogen level, and an experimental distribution exists for each standard range of hydrogen. The probability of cracking was thus determined from comparison between the predicted hardness distribution and the distribution of critical hardness for cracking, see Fig. 7.

The model developed has been shown to predict probabilities of close to 1 in 200 for threshold conditions on the TWI nomogram and between 0.15 and 0.5 for selected measured CTS test thresholds. A software tool now exists to assess the likelihood of cracking in a given component after welding, which may be used to assess the conservatism of welding procedures. Probabilistic modelling of

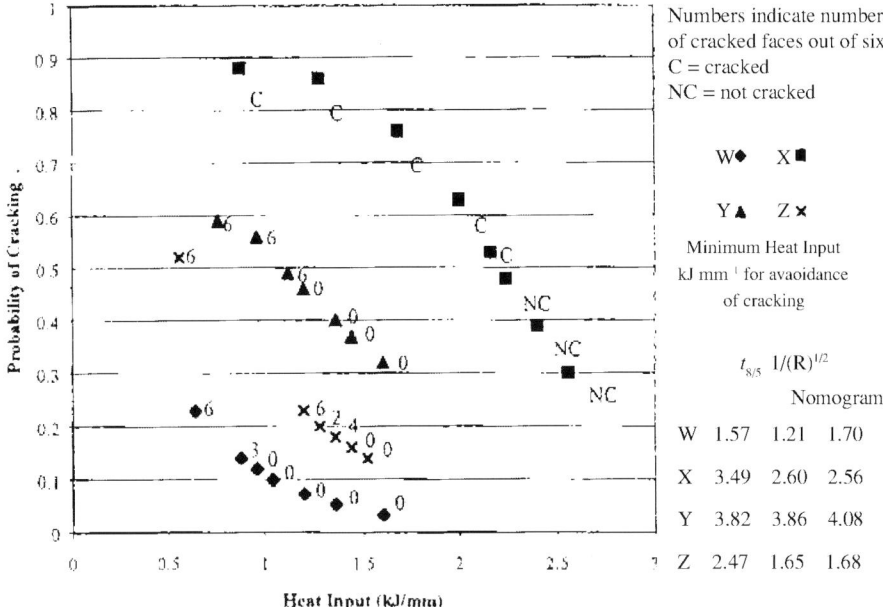

Fig. 7 Predicted probability of cracking and experimental results of four hydrogen controlled thermal severity (CTS) test series.

fabrication hydrogen cracking in C–Mn steels can be used both as a predictive and retrospective tool, for fabrication and inspection. In fabrication, the risk associated with, for example, using a lower heat input can be quantified, and inspection can benefit by knowing which components had a higher probability of cracking.

5. CONCLUDING REMARKS

Through a series of examples, an overview has been given of some recent work on structural integrity of offshore structures and components.

The deterministic assessment of the significance of flaws using published fracture mechanics remains a major part of the industry's structural integrity needs. These so-called engineering critical assessment (ECA) techniques have made great progress since the introduction of fracture mechanics in the 1950s. The methods can now be used on a routine basis for offshore plant. Software programs exist to facilitate the conduct of the analyses. A great advantage of such programs is that they enable the analyses to be repeated many times over, so that the sensitivity of the outcome to the variations in the input parameters can be studied.

The application of deterministic ECAs has been illustrated by reference to a number of case studies to show that there exist many successful examples, which can be grouped overall in the following categories:

- Analyses to show the acceptability of flaws which were outside the acceptance criteria of fabrication standards or design codes;
- Analyses to assist in material selection, design and repair;
- Analyses to demonstrate structural integrity for continued operation;
- Analyses to evaluate the need for post-weld heat treatment;
- Analyses to support the use of steels and weldments which failed Charpy toughness requirements;
- Analyses to demonstrate that operational modifications to process plant or components can be implemented safely.

The large number of successful applications of deterministic ECA to demonstrate structural integrity provides confidence that these analysis techniques are both safe and of potentially vast economic benefit to the user.

However, the offshore industry is increasingly interested in quantifying uncertainties and probabilistic methods have been developed for:

- Probability of fracture and fatigue damage with more emphasis on the quality of the material data needed;
- Inspection reliability and factors affecting it as well as further statistical work on occurrence of fabrication cracking and flaw frequency and size distributions.

REFERENCES

1. BS 7910:1999 (Incorporating Amendment No.1), *Guide on Methods of Assessing the Acceptability of Flaws in Metallic Structures*, British Standards Institution, London, 2000.
2. H. C. Cotton, 'Welded steel for offshore construction', *Proc. IMechE*, 1979, **193**(26).
3. J. D. Harrison, 'The Effect of PWHT on the Toughness of Welds for an Offshore Platform', *Proc. Conf. Performance of Offshore Structures*, Institute of Metallurgists, Autumn Review Course, Series 3, No. 7, November 1976, pp. 78–82.
4. J. D. Harrison, 'The Economics of a Fitness-for-Purpose Approach to Weld Defect Acceptance', *Proc. Conf. Fitness-for-Purpose Validation of Welded Constructions*, London, UK, November 1981, The Welding Institute, 1981.
5. R. M. Andrews, G. S. Booth and V. J. Bromley, 'Fitness-for-Purpose Assessment of Single Sided Butt Welds in Offshore Structures', *Proc. Conf. Fatigue of Welded Constructions*, Brighton, April 1987, The Welding Institute, 1987, pp. 231–239.

6. R. M. Andrews and S. S. A. Gholkar, 'Practical Analysis Method to Assess the Effects of Defects in an Offshore Structure', *Proc. Conf. Welded Structures* 90, The Welding Institute, 1990, pp. 98–108.
7. R. H. Bond, N. R. Cameron, R. E. Taylor, J. D. Harrison, M. S. Kamath, P. M. Hart and C. A. Bainbridge, 'The Buchan Alpha platform: A Comprehensive Post-Construction Fitness Review', *Proc. 2nd Int. Conf. Offshore Welded Structures*, London, November 1982.
8. J. D. Harrison, 'The Welding Institute studies and significance of Aleyaska pipeline defects', *Welding Institute Research Bulletin*, 1997, **18**(4), pp. 93–5.
9. H. I. McHenry, D. T. Read and J. A. Begley, 'Fracture mechanics analysis of pipeline girth welds', in *Elastic-Plastic Fracture*, ASTM STP668, American Society for Testing and Materials, 1979.
10. G. Barreau, E. Magne, P. Morvan, D. Tran, P. Tranter, F. Williford, G. S. Booth, P. J. Mudge and H. G. Pisarski, 'New lease of life in the 704', *Oilfield Review*, April/July 1993, **5**, pp. 4–14.
11. H. G. Pisarski, G. S. Booth, P. J. Mudge and P. Morvan, 'Life Extension for a Semi-Submersible Drilling Rig', *Proc. Conf. Materials Ageing and Component Life Extension*, Vol. II, EMAS 1995, pp. 1019–1032.
12. I. Hadley, C. I. K. Sinclair and E. Magne, 'Life Extension of Semi-Submersible Drilling Unit', *Proc. Conf. OMAE 95*, ASME, 1995.
13. I. Hadley and S. Manteghi, 'Remnant Life of Semi-Submersible Rigs', *Proc. IMechE Conf. on Remnant Life Prediction*. London, 26 November, 1997.
14. O. L. Towers, 'SCC in Welded Ammonia Vessels', *Metal Construction*, August 1984, pp. 479–485.
15. J. D. Harrison, S. J. Garwood and M. G. Dawes, 'Case Studies and Failure Prevention in the Petrochemical and Offshore Industries', *2nd Nat. Conf. on Fracture and Fracture Mechanics*, Johannesburg, 26–27 November 1984, Pergamon Press, 1984, pp. 281–296.
16. H. G. Pisarski, 'Life Extension – Ensuring Continued Pipeline Integrity', *Proc. Conf. Marine Technologies and Strategies for Mature Fields in the New Economic Era*, McDermott-ETPM, Dubai, UAE, March 1996.
17. I. Hadley, R. Phaal and S. Hurworth, 'Reliability of LPG Bullets', *IMechE Conf. Pressure Systems Operation and Risk Management*, C502/021/95, October 1995, pp. 103–112.
18. D. Noble and K. Goodwin, Private communications, 1998.
19. H. G. Pisarski, R. Phaal, I. Hadley and R. Francis, 'Integrity of Steel Pipe During Reeling', *Proc. Conf. OMAE 94*, Vol. V, ASME 1994, pp. 189–198.
20. R. Phaal, M. Z. Chen and E. Warren, 'Determination of Maximum Girth Weld Repair Lengths for Offshore Pipelay', *Proc. Conf. OMC 97*, Ravenna, 1997.
21. J. R. Still, J. B. Speck and G. R. Razmjoo, 'Integrity of FPSO Hull Structures', *The Naval Architect*, March 2000, pp. 28–36.
22. J. R. Still and J. B. Speck, 'Hull Weld Quality Critical for Offshore Oil Production Vessels', *Welding Journal*, August 2000, pp. 33–38.
23. J. R. Still and J. B. Speck, 'Early Scrutiny Yields a Healthy Hull', *Offshore Engineer*, September 2000, pp. 42–44.

24. J. B. Speck and J. R. Still, 'Making a Turn Around the Deck', *Welding and Metal Fabrication*, June 2000, pp. 19–23.

25. M. Pereira, J. B. Speck, G. R. Razmjoo and J. R. Still, 'Programme of Investigation for In-Service Damage of FPSO Hull Structures', *OMAE 2001*, Rio de Janeiro, Brazil, ASME, 2001.

26. J. Buitrago and N. Zettlemoyer, 'Fatigue Design of Critical Girth Welds for Deepwater Applications', *OMAE 1998*, Lisbon, Portugal, ASME, 1998.

27. S. J. Maddox and G. R. Razmjoo, 'Fatigue Performance of Large Girth Welded Steel Tubes', *OMAE*, Lisbon, Portugal, ASME, 1998.

28. G. R. Razmjoo, 'Fatigue Performance of TLP Tendon Girth Welds', *Sixth International Offshore and Polar Engineering Conference*, Los Angeles, 1996.

29. K. A. Macdonald, S. J. Maddox and P. J. Haagensen, 'Guidance for Fatigue Design and Assessment of Pipeline Girth Welds', *OMAE 2001*, Rio de Janeiro, Brazil, ASME, 2001.

30. J. B. Wintle, A. Muhammed and C. R. A. Schneider, 'New TWI Research on the Reliability of Welded Structures', *27th MPA Seminar*, 4–5 October 2001, Stuttgart, Germany.

31. A. Muhammed and J. B. Wintle, 'Development of an Enhanced Probabilistic Fatigue Method for Structures under Variable Amplitude Loading', *TWI Research Report* 752/2002, July 2002.

32. C. R. A. Schneider, A. Muhammed and R. M. Sanderson, 'Predicting the Remaining Lifetime of In-Service Pipelines Based on Sample Inspection Data', *Insight*, Vol. 43 (2001), 102–104, British Institute for Non-Destructive Testing, London.

33. J. B. Wintle, 'An Introduction to Risk Based Inspection', *TWI Research Report* 722/2001, February 2001.

34. J. B. Wintle, R. M. Sanderson and P. H. M. Hart, 'A Review of Methods for Determining the Frequency and Size Distribution of Welding Flaws in Steel Fabrications', *TWI Research Report* 749/2002, June 2002.

35. J. B. Wintle, R. M. Sanderson and C. R. A. Schneider, 'An Appraisal of Statistical Inspection Strategies and Inspection Updating', *TWI Research Report* 750/2002, June 2002.

36. J. B. Wintle *et al.*, 'Best Practice for Risk Based Inspection as Part of Plant Integrity Management', TWI Report on *www.hse.gsi.gov.uk*.

37. N. Bailey *et al.*, *Welding Steels Without Hydrogen Cracking*, 2nd edn, Abington Publishing, Cambridge, UK, 1993.

38. R. J. Pargeter, A. Muhammed and J. M. Nicholas, 'Probabilistic Modelling of HAZ Fabrication Hydrogen Cracking in C–Mn Steels', *TWI Research Report* 711/2000, August 2000.

CHAPTER 10

Integrity Code Applications in the Transportation Industry

Brian Dabell

nCode International, 26877 Northwestern Highway, Suite 220, Southfield, MI 48034, USA

ABSTRACT

This paper examines the importance of structural integrity in the transportation industry and outlines a number of opportunities to improve this process as well as describing the associated value propositions. The importance of addressing the gaps in our knowledge and the impact on structural integrity are discussed. It is concluded that structural integrity is not just a calculation or code; it is a process.

BACKGROUND

Quality and reliability are key expectations for both product and service providers in today's competitive business environment. These expectations can only be consistently met when the durability and performance of products and systems are well understood. This paper focuses on the importance of structural integrity and durability in transportation industries.

The makers of transportation equipment (e.g., vehicle manufacturers) and the operators of such equipment (e.g., train operators) need tools to predict equipment integrity and performance. These tools convert specific measured data into useful information to enable engineers to make good business decisions.

Examples are provided to illustrate the value of understanding product integrity to improve business performance.

STRUCTURAL INTEGRITY – ORIGINAL EQUIPMENT MANUFACTURERS (OEMs)

THE VALUE PROPOSITION – ORIGINAL EQUIPMENT MANUFACTURERS

Despite the availability of many powerful Computer Aided Engineering (CAE) tools to predict performance, the cost of failure continues to increase. Product recalls, increases in warranty costs, product launch delays and products failing in service dramatically affect company performance and consumer confidence.

Organisations that get it right can dramatically improve bottom line performance. Avoiding service failures, adding value to the product and accelerating design and development are a consequence of having the right tools, people, culture and processes in place. It is vital to recognise that the effectiveness of this combination depends on understanding the data that is required to feed the process.

Figure 1 illustrates the durability process for the transportation industry. This is a multidisciplinary process where the value is crucially dependent on the quality of key input data. External loads are applied to the transportation equipment (e.g., vehicle) and these loads are transferred through the equipment from one component to the next. The durability of each component is dictated by the loading environment experienced, the distribution of stresses and strains caused by the loading and the response of the material from which it is manufactured. As a result, the major inputs to any durability analysis are loading, component geometry and properties describing material behaviour.

Several opportunities exist to accelerate and streamline this process that can have significant benefits.

- Evaluating the durability performance of the equipment and components at the early design stage can eliminate expensive prototypes and assess more alternatives to identify the preferred solution for prototype evaluation.
- Understanding the effects of production variables in terms of material, manufacturing process effects and tolerances to allow robust designs to be created with no surprises emerging at the expensive prototype or production stages.
- By creating accelerated laboratory tests for equipment and components that simulate operational conditions, many weeks of traditional testing can be saved. Fatigue editing software to eliminate non-damaging events from loading sequences used in accelerated laboratory simulation tests provides the key.
- Rapid access to the major input data (loads, component geometry, material properties) and the streamlined flow of data through the durability process provides an opportunity to accelerate development time. In many organizations this process is either very disjointed (e.g., sometimes using different departments with no interaction) or does not exist at all. A consistent well-defined process provides many advantages as organizations strive to capture best practices, obtain more added value from expensively acquired data, transition processes to the supplier community and establish defined processes for ISO and Six Sigma initiatives.

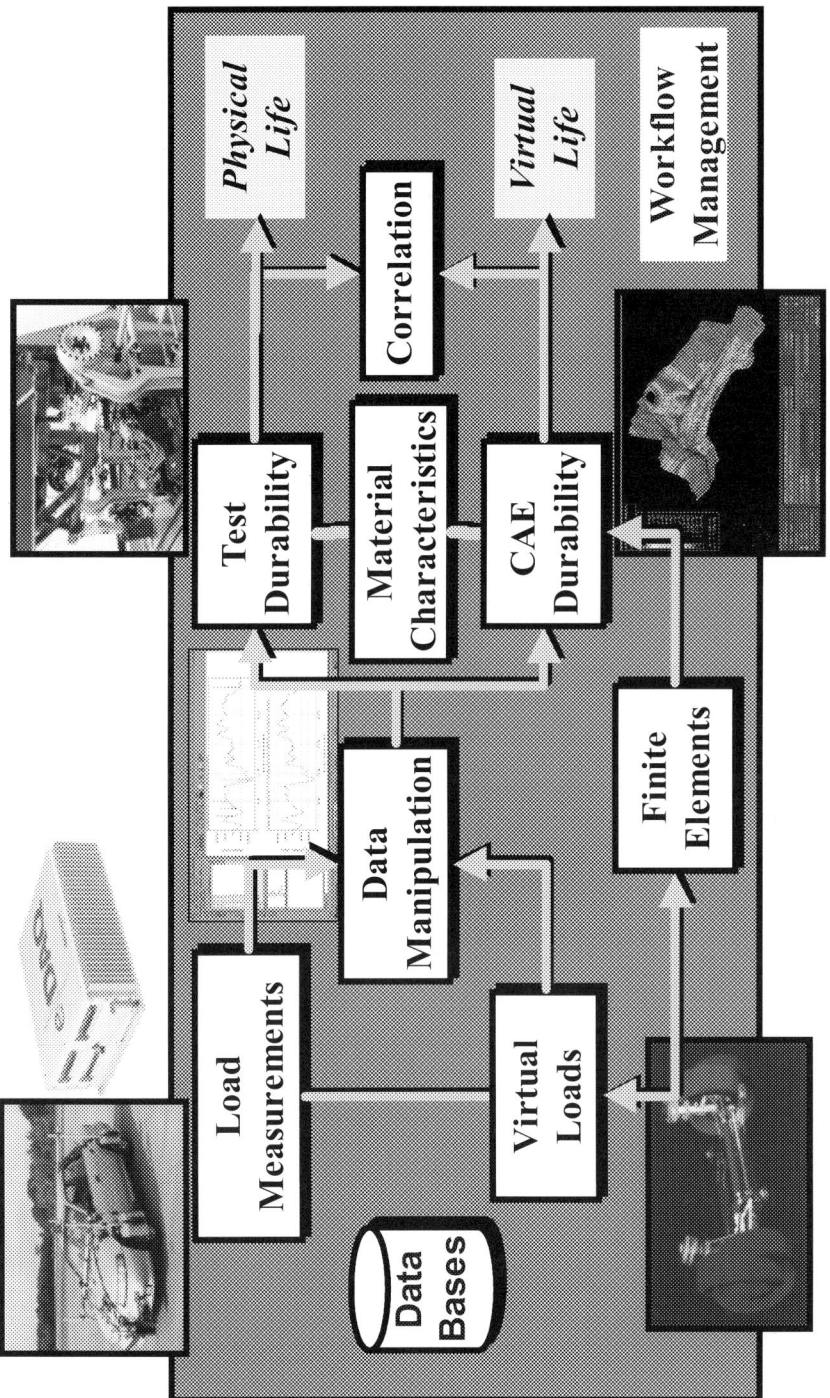

Fig. 1 The durability process.

The aforementioned illustrate some opportunities to reduce time to market. The value of this should not be underestimated by only considering direct savings: early product introduction can have a huge effect on market share and product life.

The dramatic benefits to be achieved by predicting structural integrity at an early stage of design can only be fully realised if the right data is rapidly made available when and where it is required. Data must be the core of the process and possess the intelligence to know the next stage of the process, where it is going, what format it should adopt, who needs to be informed of changes and associativity (i.e., other data, products and systems affected). This needs to be transparent to the user who will be guided through the process using best practices and can concentrate on interpreting engineering results to make decisions. Such capability is now reaping major benefits in a number of organisations.

Typical results of successfully applying the durability process are illustrated in Fig. 2. Here, CAE durability prediction software is used to understand the potential performance of a spot welded truck cab in the early stages of design. Interpretation of this information provides the design engineer with a comprehensive understanding in terms of:

- Spot welds that could potentially fail in service (the software shows the shortest predicted lives). The solution is re-orientation and/or strengthening needed in these areas.
- Spot welds that could be removed without affecting the structural integrity of the cab (the software shows the longest predicted lives).

The result was an accelerated redesign to eliminate redundant spot-welds, thereby accelerating production and reducing the investment in production tooling/spot weld stations. These opportunities represent dramatic Return on Investment (ROI) through the use of durability prediction software.

THE CHALLENGES (OEMs)

Many organisations have different departments undertaking different aspects of the durability process to understand and build structural integrity into equipment and products. Traditionally, Test and Evaluation Departments have carried out durability evaluation late in the development process using limited numbers of prototypes in limited tests to provide limited information. The 'limited' nature of the traditional approach has led to a need for conservative 'bogey testing' to overcome uncertainty. However, reality in terms of warranty costs, recalls, etc. indicates the traditional approach still has significant limitations with high associated business costs. A time consuming, prototype dominated process for

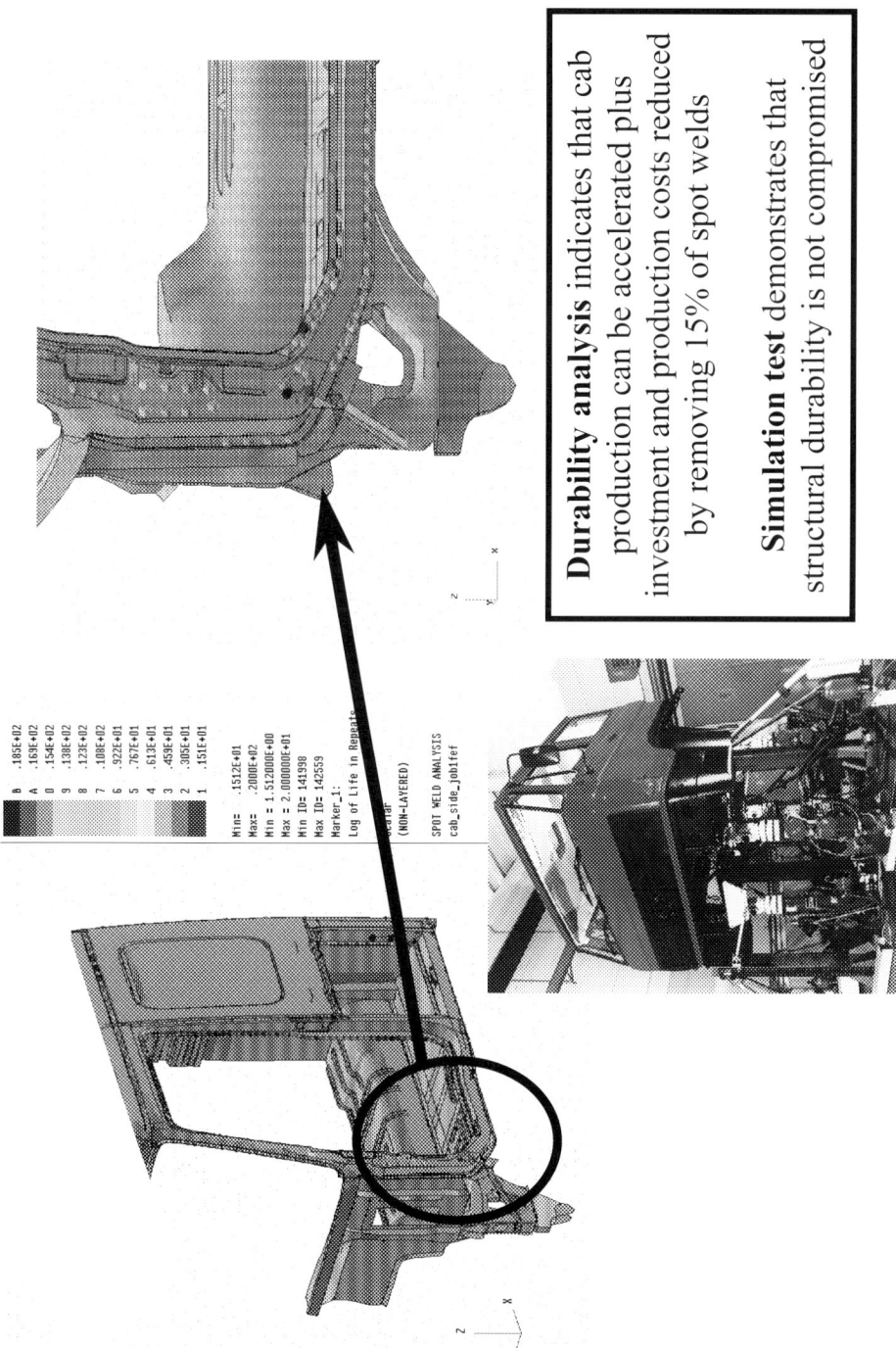

Durability analysis indicates that cab production can be accelerated plus investment and production costs reduced by removing 15% of spot welds

Simulation test demonstrates that structural durability is not compromised

Fig. 2 Value proposition: spot welded cab.

assessing structural integrity is flawed, as it cannot realistically consider the implications of production variability. Shifting the evaluation of structural integrity to the early stages of design (Design Departments) addresses many of the limitations of the traditional approach. The full benefits can only be realised when:

- Common goals exist within organisations and barriers between departments are eliminated.
- Training in the durability process and recognising the importance of understanding loads, materials and manufacturing processes is a part of the culture.
- Tools exist to manage and process the vast amount of data and information associated with the process.
- Workflow tools are created to guide the users through the durability process capturing best practices, audit trails and intellectual capital.
- Tools are appropriate for engineers rather than specialists to use and they improve productivity.
- Probabilistic approaches are used to better understand the effects of variables on structural integrity – operating environment, material variability, manufacturing tolerances, etc.
- Opportunities for improvement associated with over-design can be addressed with design and manufacturing disciplines.
- The ability exists to integrate information concerning field product operation, manufacturing and assembly capability into the durability process to maximise the value proposition.

STRUCTURAL INTEGRITY – EQUIPMENT OPERATORS

THE VALUE PROPOSITION – EQUIPMENT OPERATORS

Arguably, if the original equipment manufacturers (OEMs) get everything right then the structural integrity of the equipment should meet the needs of the user or operator. Unfortunately, this is not always the case. Some possible oversights are when the operational loading environment changes due to different customer usage, when the design loading inputs used did not anticipate all potential user operations, or when all production variables were not taken into consideration.

Whilst individual consumers who purchase products will be extremely unhappy if the product fails the impact may be very localised. However, if many consumers experience the same problem, both the manufacturers and operators will be faced with very expensive campaigns to fix the problem and potential loss of consumer confidence. In these instances the manufacturers and operators face significant costs associated with unexpected failure and the consumer is directly impacted. This unexpected failure may be avoidable by adopting the approach described earlier. Where equipment is purchased by an organisation to provide a service (e.g., train, lease car fleet) the impact of unexpected failure can have very significant business implications.

As an example, operational efficiency of organisations such as the rail industry depends on the structural integrity and reliability of the equipment: both infrastructure and vehicles.

Consider two aspects of operating a railway: vehicle reliability and track safety. The interaction between the vehicle wheels and the railway track impacts the structural integrity of both. A worn or damaged wheel will damage the track and poor track will damage the vehicles running on it. This interactive 'system' between the wheel and track must be understood.

A problem occurs when the operational loading environment for this 'system' is different to the one assumed for the design and development of the vehicle and the track. Many factors affect this such as:

- Wheel 'flats' caused by wheels slipping due to poor adhesion and brake application problems causing wheels to slide on the track. Wheel 'flats' cause very significant impact loads to the track.
- Irregular wheel wear due to operational duty cycle variability.
- Ground subsidence causing deterioration and change in the support of the track.

Thus, understanding the loading environment experienced by the wheels and the track will provide an overall benefit to predict potential problems regarding structural integrity of both vehicles and infrastructure.

On-site measurement systems (Fig. 3) exist to monitor the condition of wheels as the train passes over certain sections of track (e.g., WheelCHEX, TreadVIEW). These provide information and trends of deterioration to the train operator to allow decisions to be made for preventative predictive maintenance. The early indication of the onset of wheel damage allows cost-effective vehicle maintenance actions (e.g., wheel-set repairs) to be taken by operators before the situation becomes critical and damaging to the track and the vehicles. This avoids:

- Excessive damage to the track. Wheel 'flats' are a major cause of fatigue damage to the track and avoidance of 'flats' increases rail life and reduces the incidence of failure.
- Excessive damage to the vehicle running gear and passenger discomfort.
- Scrapping wheel-sets, since early detection of the onset of a problem enables repairs to be scheduled based on rate of deterioration.

Measure Dynamic Wheel Forces and Track Condition

Understand and predict the deterioration of track and vehicles

Track damage locations

Vehicle system damage contours

Fig. 3   Vehicle system and asset management.

In parallel, vehicle-borne systems monitor the ride of the vehicle as it travels down the track. This information can be fed back immediately to the operator who can evaluate trends of deterioration to predict when maintenance of the track will be necessary. This provides for more efficient operation and avoidance of unexpected structural failure with associated unacceptable delays in the train service provided.

The key to success is identifying the critical parameters to measure and convert the associated measured data into good information that allows straightforward decision-making. A great deal of data is generated that could be overwhelming to the operator. Therefore, it is important that any system manages the data to maximise its current and future value plus provide a simple directive to the operator regarding equipment status.

Another example relates to diesel-electric locomotive performance. Maintenance is a key operational cost and is influenced by a number of factors including:

- Current inefficient maintenance schedules that are based on hours of operation.
- Unexpected failures.
- Problems that occur when a wrong diagnosis results in the wrong parts being fixed or replaced and subsequent failure due to the original problem that was never corrected.

On-board condition monitoring equipment (Fig. 4) measuring various aspects of engine and vehicle performance can detect specific situations. When the current conditions are outside expected operating limits a message is sent immediately to the operator to enable a rapid decision to be made to address the problem based on good information and not just subjective judgement. In addition, trend information is gathered to enable predictive maintenance to prevent unscheduled failures. This provides a basis for improved maintenance efficiency and supply chain management.

THE CHALLENGES (OPERATORS)

Operators are faced with many challenges and it is essential not to add to the complexity by providing vast amounts of data without clear meaning in terms of operational impact. The data measured must provide a clear indication of operating conditions and detect key changes that will affect performance. The data must be converted into useful information and presented to the operators in a form that is straightforward to understand and enable rapid business decisions to be implemented.

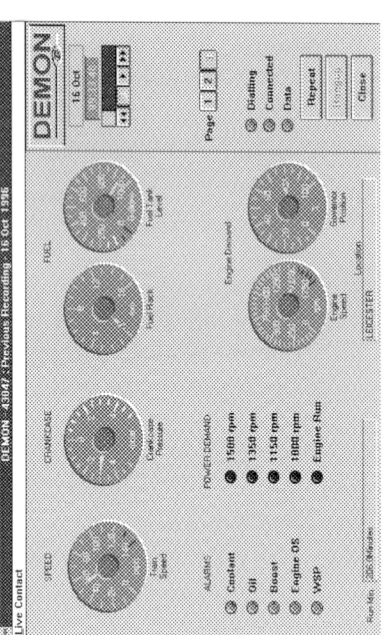

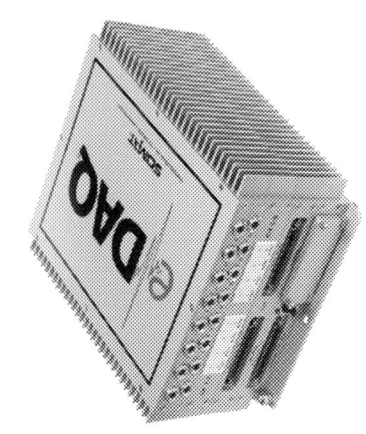

Predicting Diesel Engine Performance

Fig. 4   Vehicle systems management.

The challenges to effectively implement these initiatives include:

- Easy access to information (not data) throughout organisations for effective decision-making – clear decision trees/processes must exist.
- Common goals within and between organisations that encourages sharing of information and integrated approaches.
- Staff training in the latest methods, tools and use of information.
- Understanding the process as a whole to identify opportunities for improvement and what input data is crucial to success.
- Understanding the effects of variability and how to capture a comprehensive understanding of operational loading.
- The creation of tools that are usable by those who will benefit most from their implementation.
- Capturing the 'best practices' and experience of operators to ensure that the right information is presented in the right format to the right people at the right time to have maximum value.

SUMMARY

Overall, these examples illustrate that even in a well-regulated environment, like the railways, where structural integrity guidelines exist there are situations where the design development process fails. Any design and development process is only as good as the knowledge input into the process, as for example the expected operational loading.

Best practices can be implemented at the design and development stage but the sensitivity of the design to potential variables in operational loading and manufacturing production must be considered. Long term, unattended on-site and on-board measuring systems are invaluable in protecting valuable assets, thus ensuring structural integrity and improving operational costs and efficiency. The value and investment in such systems should increase as organisations strive to maximise the value of assets.

Understanding the true value of these tools and methods for structural integrity often lies within manufacturing, production and operational departments and not with technical specialists. The 'bridge' must be created to link what is technically possible with what is needed to manage successful manufacturing and operational businesses.

Finally, it is very important to recognise that structural integrity is not a calculation or a code; it is a process.

ACKNOWLEDGEMENTS

The author would like to thank Stacy Mills for preparing the manuscript and illustrations and Mark Pompetzki for his invaluable review.

CHAPTER 11

Current Concepts and Future Needs for the Assessment of the Integrity of Components and Structures

John Knott

The University of Birmingham, Edgbaston, Birmingham B15 2TT, UK

ABSTRACT

The paper first describes the overall aim of engineering enterprises, emphasising the role that structural integrity has to play, both in the design and construction of new plant and in the assessment of existing plant, either for continuing operation or for plant life extension. The technical basis for structural integrity assessment relates to the parameters of stress, defect size and the material's resistance to crack extension. Critical concepts in each of these areas are reviewed and suggestions for further development are made. Attention is drawn to resonance, and to secondary stresses, residual stresses, stress-relaxation and high-temperature stress/strain fields. Issues concern the ability of such stresses to continue to provide driving-force as a crack extends. Comments are made on the resolution and scheduling of non-destructive inspections, on the sources of common defects in materials, and on how fatigue life may be assessed for components which contain residual stress as a result of surface treatment. Mechanisms of monotonic and fatigue failure in metallic alloys are summarised, and attention is paid to the analysis of scatter in properties. Damage induced by in-service operation is described. The necessity to follow operational rules and maintenance schedules is emphasised throughout the paper and thoughts on future applications of condition monitoring are offered. It is recognised that the technical case is part of a broader safety culture and that the ultimate success of an enterprise incorporates a range of 'human factors'.

INTRODUCTION

The success of an engineering enterprise often relies critically on the integrity of the associated *plant* which has to deliver the primary technical *function* and, in doing this, is subjected to service *duty*. *Integrity* may be defined as the continuing ability of the plant to withstand service duty without failure throughout the design life of the plant. The enterprise is expected to achieve its aims in an

187

economically efficient manner. It is important that the financial considerations are not based only on initial start-up costs, but also include a through-life assessment. This may be affected by sociopolitical perspectives, e.g. in moves towards the lowering of greenhouse-gas emissions', or by changes in supplier-customer relationships. 'An example of the latter is the trend for airlines to require 'power-by-the-hour' rather than aero-engines as such. This forces manufacturers to re-assess their financial and technical strategies with respect to initial selling-price, maintenance costs, refits and supply of spares. Quantification of these issues necessitates inputs from through-life structural integrity analyses and the change in the customer's stance forces these to be considered in depth at the initial design stage.

Structural integrity analyses often tend to focus on the 'big issues', on those components or assemblages that are safety-critical, those whose failure would produce complete loss of primary function, leading perhaps to loss of life. Arguably, the through-life ability of plant to deliver the functional expectation of the enterprise is an equally important feature. Public confidence in the overall system is paramount. The failure of a minor component in the system may have no consequences with respect to safety, but, if it occurs frequently and its repair requires unplanned shutdown of the operating system, the loss-of-production generates, not only immediate, short-term operational losses, but also increasing loss-of-confidence in the system as a whole. This can have serious consequences for the perceived viability of the total enterprise.

In the first paper of this Symposium addressing assessment needs,[1] attention was drawn to the 'bath-tub' curve, expressing failure rates as a function of time in service. We are led to expect 'birth pangs' or 'teething troubles' at the beginning of life, and general 'wear and tear' or 'rusty joints' as a result of 'ageing'. There may also be rather more unexpected 'mid-life crises'. Guarding against each of these different aspects may involve different strategies and techniques. More precisely, in continuing to deliver the primary function, plant components must not fail within the design lifetime, or the scheduled replacement period, under the *service duty* implied by the function. The duty may include mechanical factors, such as applied stress, secondary stress, or thermally induced strain, often varying with time; chemical factors, such as corrosion or oxidation, again time-dependent; and physical factors, such as high or low operational temperatures. In some circumstances, the duty experienced may be more extreme than that anticipated. The properties of the materials used to construct components may also degrade with time. The first step is to assess the effects on through-life integrity of the normal duty as part of the original design procedure.

A number of inputs may contribute to the derivation of the anticipated duty. One is *information*, based on past experience. If a newly-derived duty cycle does not bear (extrapolated) similarities to what has been experienced previously, critical re-examination of assumptions should be a mandatory requirement. In many cases, past experience is encapsulated as *codified model duty cycles* (e.g. CARLOS for automobiles, FALSTAFF for aeroplanes) and further codification is seen in *operational codes*, specifying limits on factors such as the pressure-temperature 'envelope' that is permissible when bringing plant up to full power. To attempt to avoid 'mid-life crises', a second important input is *intelligence*, relating to the extremes of natural events in the geographical region where the plant has to operate. Such extremes might be the genuine magnitude of the '100-year wave' in the North Sea, the chances of exceeding the design-basis earthquake in Japan or Turkey, or the likelihood, in Bangladesh, that a river may flood upstream and affect plant downstream, by taking an unexpected course. The consequences of such 'knock-on' events can be startling. Intelligence *might* have been able to predict that Mount St Helens would erupt, but the effect of the ingestion of its volcanic dust by aero-engines was not immediately foreseen as a consequence. The intelligence must be coupled with *imaginative extrapolation*. This may have to be exercised on a broad front. The enterprise is expected to operate over a period of time, and during that time, public expectations may change. Road bridges provide an example. Here, an original design life may have been planned as 120 years, based on a traffic loading spectrum for which the high loads corresponded to maximum axle weights on lorries of 32 tonnes, and on a given split of freight traffic between road and rail. Throughout the last 20 years, the axle weights have been allowed to increase to 38 tonnes and there is an increasing amount of freight traffic being carried by roads. There are now proposals to allow a further increase in axle weights, to 44 tonnes. The consequence of such factors is that the operating life of a bridge will be shorter than that assumed in the original design. It is of interest to query whether such trends could ever have been anticipated at the time when the bridges were built and allowed for by an initial degree of over-design. This would clearly have been regarded as 'uneconomic' in terms of start-up costs, but would be justified by a through-life analysis, because refurbishment or rebuild costs would be reduced. Such considerations imply a long-term view, which, in general, will be associated only with projects sponsored by stable organisations: Government, nationalised industry, or very large, well-established, private industry.

An issue of current interest, relating to public expectations, is the looming 'carbon tax', encouraging car designers to reduce weight, by using aluminium or magnesium alloys to replace steel components. Structural integrity issues relate to the lifing consequences of moving to light alloys for structural purposes: in particular, to the properties of joints. The overall economic case rests on the initially high start-up costs (because more expensive materials and processes are being used) vs. the through-life costs of lower fuel consumption, combined with an increasing level of Government taxation. The car-purchasing public (both 'fleet' purchasers and private individuals) then need to be convinced (probably by fiscal measures) that it will be advantageous to buy initially more expensive cars. Social factors, not necessarily considered by the designer, assume importance: for what period of time is the car going to be used before it is sold; what will its second-hand value be? The success of the 'engineering enterprise' here clearly does not rest on structural integrity factors alone, although these factors form an integral part of the overall technical assessment.

To derive an informed perspective on any threat imposed by duty, it is necessary to define an *agreed descriptor* of what part of the duty should be regarded as potentially damaging and how this might be quantified through measurable parameters. For mechanical fatigue loading, the descriptor might be the frequency of occurrence of a particularly high load-amplitude; it might involve 'rain-flow' counting; it might require the generation of a 'power spectrum' by effecting a transform from the amplitude-time domain to an amplitude-frequency domain. *Resonance* may amplify the effect of quite modest duty amplitudes and resonant peaks therefore need to be established for any dynamic system. Given the agreed descriptor, the informed perspective on a probability of failure is given by comparing the frequency of occurrence of damaging duty with an *agreed data base* relating to material failure. This might simply comprise 'S–N' data for smooth or notched test-pieces, or it may be data for joints or sub-assemblies. Ideally, the information is sufficient to be described by a credible statistical distribution, so that probabilities can be compared to derive the overall failure probability. Design Curves, such as those produced by TWI for welded joints, may already be based on an agreed level of probability, e.g. (a linearised version of) 2 s.d. below the mean.

It is then necessary to explore the *consequences* of any failure mechanism and relate these to the *risk*. The consequences and risk could be purely financial. A large loss-of-coolant accident (LOCA) in a PWR pressure-vessel may be guarded against by the operation of the emergency core-cooling system (ECCS), which floods the vessel with cold water. This leads to a *safe shutdown*: 'safe', in the sense that the possibility of exposure to a dose of more than 0.1 mSv at the boundary fence is less than 10^{-7}, but a 'shutdown', nevertheless, which means that it is highly unlikely that the plant would ever be permitted to operate in the future. It is not usual to express risk in terms of the probability of exposure to noxious emissions and there is a tendency to work in terms of the risk of death to individuals. Figure 1 shows a representation of the Health and Safety Executive's (HSE) recent (1999) proposals.[2] These are expressed as a graph of F(N) vs. N, where F(N) is the probability of a failure which kills N individuals. The HSE proposal (for discussion) is that, for engineering plant, the annual probability of occurrence of an event that causes the death of 50 individuals should be less than 1 in 5000.

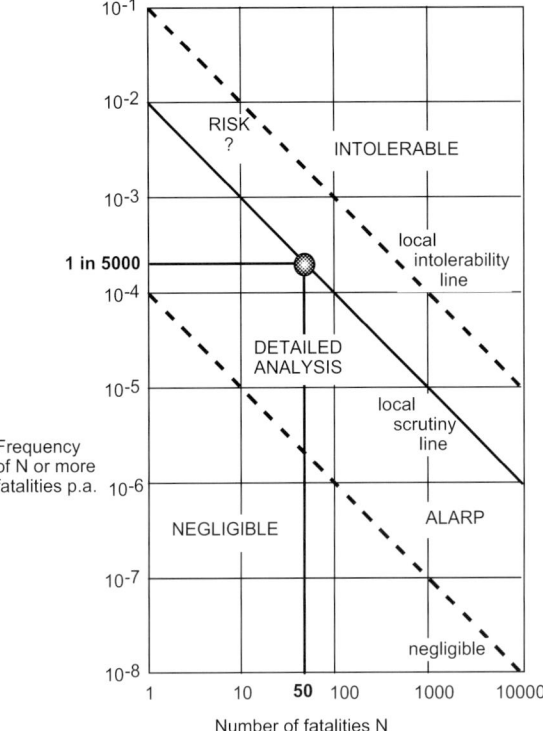

Fig. 1 The F/N curve from the HSE Document 'Reducing Risks, Protecting People' (June 1999).

INITIAL BUILD AND CONSTRUCTION

A simplified representation of the constraints that may have to be faced when contemplating the design of plant which has to deliver the aim of a *new* enterprise is given in Fig. 2. Three factors are shown: *efficiency* (associated with high stresses on components), represented by the £ sign; *lifetime*, represented by the hourglass; and *safety*, represented by the life-belt. Keeping all three in balance requires the training and experience of a skilful performer. The initial design process should no longer be regarded as an isolated activity. Formerly, it might have involved broad (architect-like) concept designs, turned into 'virtual reality' in the stress office. Today, *design* should be far more involved in the *totality of the enterprise*, and Fig. 2 is very much a simplification. There is a plethora of, often competing, input parameters: satisfying the prime function; guaranteeing the design life and addressing the safety issues that this implies; minimising the initial costs of

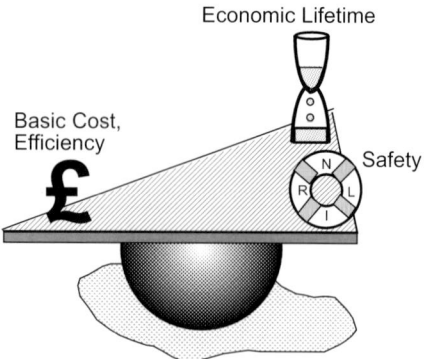

Fig. 2   The 'Balancing Act' between Efficiency, Economic Lifetime and Safety.

manufacturing, assembly and materials; minimising the through-life costs of maintenance and component refurbishment or replacement; meeting environmental concerns on noise levels, greenhouse-gas emissions or effluents. This is a balancing-act in many dimensions. As a broad generalisation, it may be that future designs will tend to be a little less adventurous in terms of initial promises, if the consequences are trouble-free, 'sweeter' running throughout the design life. In summary, *'less* BRAVURA, *more* DOLCE VITA'.

New concepts can bring problems arising from dealing with the unfamiliar. The box-girder bridge sections of the early 1970s were introduced to reduce the labour-intensive assembly methods associated with truss frameworks, but lack of proper understanding of the buckling behaviour of thin plate in trapezoidal cross-sections under compressive loading led to a number of failures during the erection process. The point here is that analysis of the designers' drawings of the *finished* bridges (the 'virtual reality') would have shown them to be perfectly satisfactory. It was the occurrence of unbalanced loads, leading to excessive bending moments during the *construction* phase, that caused failure to occur. Concrete bridges and buildings have also collapsed during the construction phase, as a result of the inadequacies of the 'false-work' used to support the concrete during pouring and setting. These examples show how important it is for the *rigorous processes* associated with the original design and life-assessment analyses to be applied throughout the *construction, fabrication and manufacturing stages.*

A relatively new area of concern is that of the modern 'death-defying' *rides* which are now appearing in fairgrounds and amusement parks, bearing names such as *Oblivion* or *Nemesis*. The *Traumatizer*, at Southport, stands 115 feet off the ground, has a top speed of 53 mph and lifts passengers to 100 feet before releasing them into a 'rollover corkscrew and a sidewinder half-loop'. It records positive g-forces of 4.5 (g-forces of 6 or more cause black-outs and nose-bleeds). The innocuous-sounding *Pepsi Max – the Big One* has a 235 feet drop followed by 'an 85 mph hurtle through a mile of twist and turns'. The function of such engineering enterprises is to produce *thrills*, but it is equally necessary to avoid *spills*. The newspaper *US Today* (for Fri/Sat/Sun 7–9 April 2000) contained a front-page article (and articles giving State-by-State information) on the situation in the USA. Between 1987 and 1999, there were approximately four deaths per annum, with six deaths in 1999. *The Times* on 29 May 2000 reported on two deaths in different fairground incidents over the Spring Bank Holiday weekend. The first, which killed a 28-year-old woman and injured two men, resulted from a chair becoming detached from a high-speed *Super Trooper* ride (at the Shepherd's Bush fair) and flying through the air some 35–40 feet off the ground. The second occurred at Redruth, Cornwall, when a 12-year-old girl was hurled 35 feet to the ground from a *Top Spin* ride. The General Secretary of the Showmen's Guild gave figures of 'more than 300 million rides per annum, with about 90 accidents being reported – there have been no fatalities for a couple of years'. The HSE said that it had served 26 prohibition notices on operators in 1999, demanding that rides be closed until safety work was carried out. Five operators had been prosecuted for negligence. *The Mirror*, 24 July 2001 (p. 15) reported that 'forty children were left hanging upside down yesterday when their white-knuckle ride . . . *Rameses Revenge* broke down . . . it was more than an hour before they could be freed'. *The Times*, 1 August 2001 (p. 13) reported on 'passengers trapped for up to five hours on a ride which sheered off its spindle . . . in Dalton Town, Michigan, injuring more than two dozen people'. Such examples point up the need for increasingly close attention to be paid to initial design, to dismantling and re-assembly procedures (for travelling fairgrounds) and to maintenance schedules.

In parallel with the modelling and codification of 'service' duty, the definition of which is now extended to include detailed consideration of the construction phase, it is important to draw attention to the need for sets of *operational rules* and *maintenance schedules* to ensure that, first, the plant is operated in the manner anticipated in the original lifing assessments; second, that regular maintenance checks are made on individual components to ensure that they continue to operate as intended. Simple examples may be found in the operation of a private car. The rev. counter has a 'red' forbidden region to prevent attempts to work the engine too hard; regular servicing ensures that brake pads are checked, cables are greased, oil-filters are changed. Warranty agreements provide financial incentives to carry out regular maintenance: the warranty is rendered void if the stamps are not present in the Owner's Handbook to show that servicing was indeed carried out at the recommended intervals. Departures from recommended practices in the operation of large-scale industrial plant and systems (e.g. Flixborough, Chernobyl, the Paddington rail crash) have led to horrendous consequences and it is of *vital importance* to any company to develop a sufficiently *strong safety culture* to ensure that such practices, tedious though they may appear to be, are followed conscientiously, with rewards given to operatives who report on unusual features, such as leaking valves or changes in the noise/vibration characteristics of pumps. Such observations and reporting form the first line of defence against 'mid-life crises'. An important assumption is that the organisation has the motivation and competence to treat such information in a serious and professional manner.

The considerations above have been concerned primarily with the design of new plant and with the need to adhere to appropriate operation procedures and maintenance schedules. Often, structural integrity assessments have to be made on plant which is either already part-way through life, or has actually reached the end of its original design, but is being assessed for extended operation. Factors here are that original information may not be complete, crack-like defects may have been detected and the material properties may have changed during operation. It is for such applications that the techniques of fracture mechanics in the broadest sense have proved so useful. The following section summarises some of the key aspects associated with fracture mechanics assessments.

THE TECHNICAL BASIS FOR ASSESSMENTS

Three parameters are linked together: the *stress*, or 'driving force for crack extension'; the *size* of any crack-like *defect* that may be present; and the material's *resistance to crack extension*, either by fast fracture or by sub-critical crack growth (which may eventually lead to a critical condition). A failure mechanism which is an alternative to fast fracture is failure of the uncracked ligament by *plastic collapse*. For thin-walled components (perhaps containing defects) it is also necessary to assess modes of *buckling* (as in the box-girder bridges during construction, in submarines during operation, or the 'Saturn' rocket on its launching gantry) and Mode III ductile *ripping* (aircraft fuselage, rocket motor-cases).

Elastic stress analyses for cracked bodies are available for a wide range of geometries and loading conditions. For analytical solutions, 'weight function' techniques give a range of flexibility, to treat, for example, the effects of residual stress distributions on crack-tip stress-intensity factors. Elastic/plastic stress distributions have been calculated for conditions of 'small-scale' yielding in cracked bodies, and for extensive plasticity (up to and beyond general yield) in selected notch-bend geometries, but the more popular route has been to assume 'power-law hardening' material properties and to calculate J-integral values for extensive plasticity. Specified values of J or crack-tip opening displacement (CTOD), δ, may be measured in generally-yielded testpieces to characterise a particular event, such as the onset of fast cleavage fracture or 0.2 mm of ductile crack extension. To assess the likelihood of failure in a structure, these values are converted to equivalent values of stress intensity factor, K, using the appropriate relationship for plane strain:

$$K^2 = EJ/(1 - \nu^2) \tag{1}$$

$$K^2 = 2E\sigma_Y\delta/(1 - \nu^2) \tag{2}$$

where E is Young's modulus, ν is Poisson's ratio and σ_Y is the yield stress. The characterising value of K, K_{Ic}, K_J, or $K_{0.2}$ is then taken as an appropriate measure of the material's resistance to crack extension, K_{mat}.

Assessment is carried out with respect to the *failure assessment diagram* (FAD), which incorporates both the possibility of failing by fast fracture and that of failing by plastic collapse. The diagram has an ordinate, K_r, which is the ratio of the 'applied K', K_{app} (generated by the action of primary and secondary stresses on any defect present) to the material's 'resistance', K_{mat}. Hence, $K_r = K_{app}/K_{mat}$. The abscissa, L_r, is the ratio of the applied load to the collapse load. The present form of the FAD bases L_r on the yield strength, but the failure locus continues beyond 'general yield', recognising that the material's work-hardening capacity can prevent collapse. Different 'cut-off' values are ascribed for different materials. A schematic FAD is shown in Fig. 3. Assessment is carried out, for a defect of interest, by calculating the appropriate values of L_r and K_r and locating the position of the point implied by these co-ordinates. If the point lies within the failure locus, the situation is deemed to be 'safe': if it lies outside, failure is predicted. The

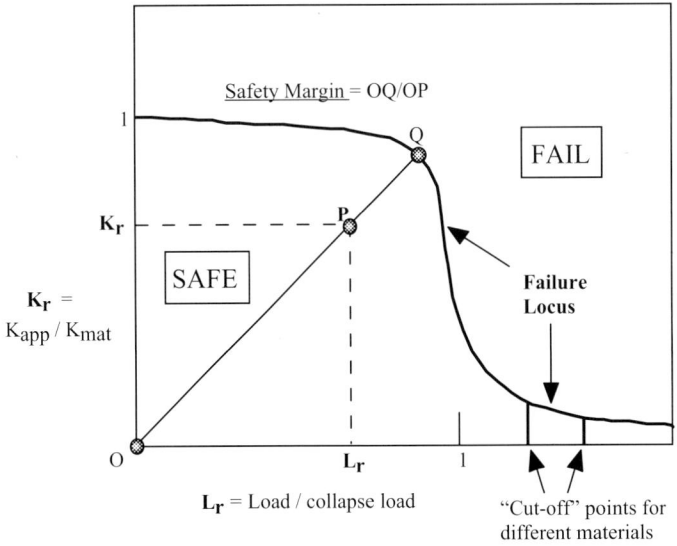

Fig. 3   Schematic Failure Assessment Diagram.

margin of safety may be expressed in terms of the (inverse of the) ratio of the distance from the origin to the assessment point, OP, to the distance at which an extension of this line cuts the failure locus, OQ: i.e. the safety margin = OQ/OP (see Fig. 3).

Crack sizes can increase during service operation, as a result of *sub-critical growth mechanisms* such as fatigue, stress-corrosion, corrosion-fatigue or creep. To assess whether a defect is likely to lead to premature failure before the next inspection period, or indeed to decide what the inspection period needs to be (and whether this is economically viable), it is necessary to input knowledge of sub-critical crack *growth rates*, to calculate the size to which the defect will have grown after a specified period of time and to reassess the situation, using the FAD, for the larger defect size. The confidence associated with such calculations is increased if the defect growth can be followed from start-of-life and if stress spectra and crack growth-rate data are well characterised; often, however, such knowledge is incomplete and rather large safety factors have to be assumed. The following sections make specific comments on stresses, defects and material properties to draw attention to some of the main issues.

STRESSES

In initial design, one of the most important features for dynamic loading is to ensure that *resonant peaks* are identified, and that any *hot-spots* in the system are located spatially. This can be determined experimentally if a sub-assembly, such as a car's suspension, is subjected to road loadings in the testhouse, via track recordings played back through servo-hydraulic actuators, and if *thermal emission techniques*, such as SPATE, are used to scan the vibrating system to detect heat produced by plastic deformation. Alternatively, the system could be coated with heat-sensitive paint/polymer before testing to indicate where attention should be concentrated. For large structures, such as bridges, use can be made of models in wind-tunnels, but other loadings, from tides or traffic need also to be included, individually and in combination. The 'wobbliness' experienced by some of the first pedestrians to walk across the new Thames Millennium footbridge was attributed by the designers to the fact that 'people were walking in step'. The bridge was closed within three days to allow extra *damping* and *stiffening* measures to be incorporated. The lack of such measures in the initial design is perhaps surprising and it is of some interest to recall comments in the first two pages of Nevile Shute's *No Highway*, written in 1948,[3] with reference to Mr Fox-Martin and Miss Bucklin (sic): '. . . they had got the unstabilised, eccentrically loaded strut just about buttoned up, regardless of the fact that unstabilised struts are very rare today in any aircraft structure'. It is not only with respect to the Millennium bridge that these words have a particular 'resonance': the failure of the hull of the gigantic *Team Philips* catamaran in 2000 can be attributed to roughly similar causes.

Secondary stresses may be produced by thermal means, either as a result of thermal gradients arising from the service duty, or as internal, *residual* stresses, generated when different parts of a structure cool through different temperature ranges or possess different thermal expansion coefficients, when cooling through a temperature range. The operation of emergency core cooling, following a loss-of-coolant accident (LOCA) in a PWR pressure vessel (running at ∼290°C) floods the inside of the vessel with cold water. The inner wall of the vessel then tries to contract and tensile surface stresses are generated. A fusion weld cools through a larger range than the components which it joins: it therefore attempts to contract by a greater amount and generates tensile stress. The severe thermal cycle to which the bore of a rifled gun-barrel is exposed during the first firing generates stress (as a result of thermal gradients and volume changes associated with phase transformations) sufficient to produce 'heat checking' cracks at the root of every rifling groove. The cladding of ferritic material by austenitic steel is effected by a weld-like deposition, and the stresses are influenced not only by the temperature range, but also by the fact that ferritic and austenitic steels have different thermal expansion coefficients. The French have experienced 'under-clad cracking' in PWR pressure vessels as a result of the combination of such stresses and embrittled material.

A point of interest is the extent to which monotonic *thermal fields* continue to provide 'driving force' for crack extension as a crack grows longer. Even for brittle ceramic materials, the use of chevron-notched test-pieces which generate steeply decreasing stress-fields as a crack extends are able to produce semi-stable crack extension, if not arrest. For the PWR LOCA, the minimum fracture toughness that would be contemplated would be over $40\,\mathrm{MPam}^{0.5}$. In the absence of further primary stresses being applied, the cracks are likely to arrest. The French 'under-clad cracks' were only some 5–8 mm in length, compared with a wall thickness of order 150–200 mm. The problems associated with a simple LOCA were originally over-emphasised, not fully taking account of the fact that, as a crack extends, *both* the thermal stress 'driving force' becomes less effective, *and* the crack grows into a warmer region, where the fracture toughness of the material is significantly higher. A given fracture toughness implies a given displacement for each increment of crack growth. The extent to which this displacement might relax the residual stress field, and how this is allowed for in the FAD-based assessment is a subject needing further detailed exploration.

It is necessary also to assess the possibility of growth of any defects initiated by thermal stress. The gun-barrel provides a good example.[4] Here, the length of an initial defect is similar in size to the depth of a rifling groove (2–3 mm) and the barrel is subjected to a sequence of closely similar pressure cycles for each firing of the gun. Integrations of the fatigue crack growth rate can be used to give highly plausible figures for the life of the gun. In particular, it can be shown that the life is affected quite strongly by the final fracture condition, i.e. by the fracture toughness of the gun-barrel material. In contrast, the number and magnitude of the pressurisation cycles throughout the design life of a PWR pressure-vessel are not predicted to lead to crack growth of any extent that would cause concern.

The above cases relate to the generation of *tensile* residual stresses and to defect formation. Much design against fatigue in general engineering practice makes use of techniques to generate *compressive* residual stresses which help to retard the growth of fatigue cracks. Rifled gun-barrels are usually subjected to an auto-frettage treatment, which cold-expands the bore and gives rise to compressive 'clamp-back' stress when the means of cold-expansion is removed. Engine components may be case-carburised, case-nitrided or induction-hardened. Many structures are peened, using shot, glass balls or hammers; lugs on aircraft may be 'ballised' by forcing an oversized ball-bearing through the lug. The mechanical effect of these treatments is to generate a distribution of compressive residual stress, which acts in opposition to the opening effect of an applied tensile stress and hence retards the early stages of crack growth. The stresses can be measured using X-ray or neutron diffraction and analysis, perhaps making use of weight functions, can be used to derive a 'negative K' which can be subtracted from the applied 'positive K' to obtain the effective stress intensity factor which controls crack growth. The effect is monotonic in sign and is less effective if the mean stress in the fatigue cycle is high. There is still a need to agree the optimum experimental and analytical techniques to address these issues.

Both stress-relief heat-treatments and stress-relaxation processes may be experienced by engineering plant. In *stress-relief*, the aim is to reduce the level of residual stress after welding. Thermally, this is achieved by heating the welded joint slowly to a temperature of order 650°C, at which it is held, usually for less than 24 h, before the joint is allowed to cool to room temperature. The intention is to reduce the stress level by plastic flow, since the uniaxial yield strength decreases markedly with increase in temperature, but cracking may be induced in susceptible steels.[5] *Stress-relaxation* is associated with plant operating for substantial periods of time at temperatures sufficiently high for the material to deform by *creep*. Under steady primary loads, 'forward creep' is experienced, with the secondary *creep-rate* exhibiting a power-law dependence on stress. Fixed-displacement boundary conditions, experienced when the main component of stress is generated thermally, allow stresses to relax as elastic strain is replaced by creep strain. In some configurations, there may be significant 'elastic follow-up', allowing the creeping region to accumulate large strains, hence contributing to *damage*. In materials possessing low creep ductility, damage accumulation can lead to creep fracture. Modern trends are to express damage as the ratio of accumulated strain to creep ductility, rather than as a fraction of the time to rupture.[6] Under cyclic operation, the new strain cycles may regenerate high stress levels, if a previously relaxed (and softened) region becomes work-hardened. This would produce a strong creep–fatigue interaction. The R5 'Assessment Procedure for the High-Temperature Response of Structures' contains guidance on how to treat these issues.

Similar factors contribute to the conditions existing in a 'creeping' enclave at the tip of a crack-like defect. Globally, the conditions are elastic and could be under either force control or displacement (total strain) control. For linear elastic or power-law hardening non-linear elastic material (*strain* dependent on stress raised to a power) the energy-*rate* ('rate' with respect to crack length) may be expressed in terms of G or J, with $K^2/E' = G$ being a formal identity for linear elastic behaviour, and $K^2/E' = J$ assumed for small-scale yielding as a characterising parameter in energy 'rate' form. The term E' is E in plane stress or $E/(1 - \nu^2)$ in plane strain. For a power-law dependence of *strain-rate* ('rate' now with respect to time) on stress, the *strain-rate fields* are characterised by the parameter C^* which has an algebraic form similar to that of J. If the boundary of the enclave is taken as constant total strain, and stress relaxation occurs as a function of time, an initially high value of C^* gradually decreases until it attains a steady-state value, calculable from the reference stress. The crack-tip *strain* field can also be calculated as a function of time and when the accumulated strain becomes equal to the creep ductility (in a multi-axial stress state) the crack advances. There may be a degree of elastic follow-up from the region surrounding the creeping enclave, but a less clear point is whether, for a macroscopically force-controlled system, the stresses in the enclave relax or the region continues to deform in (accelerating) forward creep.

DEFECTS

The issues here are concerned with the detection and sizing of defects; with the ways in which defects may be introduced in materials processing, in the fabrication of assemblies, or by unexpected abuse during service; and with the 'detective work' carried-out on failed components to characterise the defects that have led to failure. Three main strategies are employed to limit the sizes of defects in structures: the 'proof-test', non-destructive inspection, NDI, and 'process control'.

In the *proof-test*, applied to static structures, or the equivalent *overspeed test*, applied to rotating components, the part, before entering service, is subjected to stresses, typically some 20% higher than the design stress, i.e. if the design stress is two-thirds of the yield strength, the proof-test stress is 80% of the yield strength. The primary function of the test is to ensure that the part entering service can withstand loads higher than the anticipated service duty without failure. This implies that there is an upper limit to the size of defect present at start of life (also that the material has adequate toughness). A second feature is that, when the part is unloaded after proof-testing, the plastic zones developed at the tip of any defect present exert compressive residual stresses on the crack-tip region, so that the service load has to be increased (by 20% or more) to produce further forward plasticity which might initiate new micro-cracks or expand micro-voids. The proof test provides a margin for possible sub-critical crack growth in service, but this may not be large. Consider a high pressure steel tube, made from material of yield strength 1200 MPa and fracture toughness 120 MPam$^{0.5}$. A design stress of two-thirds yield is 800 MPa and the proof-test stress is 960 MPa. Taking the generic form for fracture toughness as $K = \sigma(\pi a)^{0.5}$, the values of a become approx. 7 mm and 5 mm respectively. The proof test 'proves' that there is no defect greater than 5 mm present at start-of-life, but the critical defect is only 7 mm. This gives little margin for sub-critical crack growth and other means of assessing and controlling defects are needed.

A second measure is to perform non-destructive inspection: during manufacture, at start-of-life, and at periods throughout the operation of plant. There are specialised techniques available for detecting surface-breaking defects, e.g. fluorescent dyes and magnetic inks, but the two techniques used commonly to detect buried defects are *radiography* (using high-intensity X-rays or γ-rays from radioactive sources such as Co^{60}) and ultrasonics (perhaps coupled with acoustic-emission, AE, passive monitors to locate noisy regions during the proof test). Radiography is best suited for the detection of volumetric defects, but is far less sensitive than ultrasonics for narrow, crack-like defects.[7] Consequently, attention is paid here to ultrasonic methods.

The ultrasound is generated by a piezo-electric transducer (the *probe*). For steel structures, the probe is usually 25 mm diameter and operates at a frequency of 2 MHz, equating to a wavelength of 3 mm. For extremely clean, nickel-base superalloy gas turbine discs, it is possible to contemplate a frequency of 50 MHz, corresponding to a wavelength of 0.15 mm. The most commonly used technique is that of 'pulse-echo', although 'pitch-and-catch' configurations may also be employed. In a 'solid' section, a pulse of ultrasound travels through the thickness until it encounters the 'back wall' where it experiences a dramatic change in acoustic impedance, $Z = \rho c$, where ρ is the density of the medium and c is the speed of ultrasound in the medium. Almost all the incident intensity is reflected and is detected by the probe acting in receiver mode. The reflected intensity is displayed on a CRO vs. the time base and the time taken to receive the back-wall 'echo' can be equated to the section thickness. If a crack-like defect is present in the path of the beam, it appears as a reflected intensity at a time corresponding to the depth of the defect below the surface. This is referred to as an A-scan. The lateral dimensions of the defect are determined by rastering the probe across the surface and recording the signal intensity. The edge is defined as the position at which the reflected intensity decreases to a fixed fraction (usually $-20\,dB$) of the maximum value. This 'threshold' can then be used to produce automated displays, which record whenever the threshold is exceeded. The B-scan plots depth of defect vs. probe position on a line scan; the C-scan records exceedance as a function of the Cartesian co-ordinates of the probe, i.e. a plan view.

The limits to resolution may be appreciated by analogy with optical reflection/diffraction theory, treating the defects as small (smooth or irregular) mirrors. An irregular surface reflects only part of the beam in a specular manner and can give rise to unexpected intensity peaks elsewhere. This may produce difficulties in interpretation if an array of transducers is being employed. As the wavelength of the ultrasound becomes comparable with the size of the reflecting 'aperture', strong local diffraction effects are set up and the directed nature of the reflected beam is lost. The resolution may be taken to be roughly the wavelength of the ultrasound used: approx. 3 mm in steel. If the defects are non-metallic inclusions rather than air-filled cracks, their acoustic impedances are a significant fraction of that of the steel matrix and the reflected intensity is reduced. A 'dirty' steel, containing many inclusions, produces strong attenuation. For wavelengths comparable with the steel's grain size, further attenuation can occur as the beam of ultrasound passes from grain to grain, because the elastic constants of the iron lattice vary with crystallographic orientation and both internal refractions and total internal reflections occur.

Once defects have been detected, located and sized, it is necessary to assess their importance with respect to structural integrity and it is here that fracture mechanics has an important role to play. Some codes insist that 'any detected cracks must be removed and weld repair effected'. Following such a mandate, it is arguable that, for thick, steel structures not susceptible to sub-critical crack growth, an improvement in ultrasonic resolution can be a self-defeating exercise. Fracture mechanics calculations are unlikely to show a major threat from a crack of size 3 mm, but removal and repair (which may not be fully stress-relieved) could leave the material in a worse condition than it was in originally. Such comments do not apply if the service duty induces sub-critical crack growth: it is possible to rationalise the TWI fatigue design curves for welded joints in terms of the growth of defects initially some 3 mm in size.[8] Here, improvements to performance will be achieved, not by improving ultrasonic resolution, but by quality control during welding.

A second use of fracture mechanics is to help set appropriate through-life inspection intervals. Figure 4 shows a schematic graph of crack length, a, vs. number of cycles, N, for the fatigue loading of a 'high-quality' component such as an aero-engine turbine disc, which might have a design life of 10 000 major (take-off/landing) flight cycles. There are several crack lengths of interest. The first, a_o, is the original size present in the component. The second, a_f, is the final size, which leads to failure by fast fracture or plastic collapse. The third indicates the 'highly reliable' defect sizing capability, a^*, in this case for surface-breaking cracks. It is postulated that a suitable 'first' service inspection period (following inspection during manufacture) might be set at one-third of the design life. At low stress-amplitude, this can be met with a sensible safety margin, but, as stress levels are increased, the safety margin becomes seriously eroded, because fatigue-crack growth-rates depend on the third or fourth power of stress. Inspection may involve dismantling the engine, significantly increasing through-life costs. The frequencies of inspections necessary to preserve safety margins need to be assessed and incorporated into the initial design and costing processes.[9] In the future, growth-rate data may be so well-established that it will become practicable to use in-flight sensors to record the duty experienced by each individual aircraft and to analyse this information, *either*, to decide the date of the next inspection, *or*, if inspection periods are already scheduled, to alter the routes flown by a given aircraft, so that its accumulated duty is appropriate for the planned inspection period.

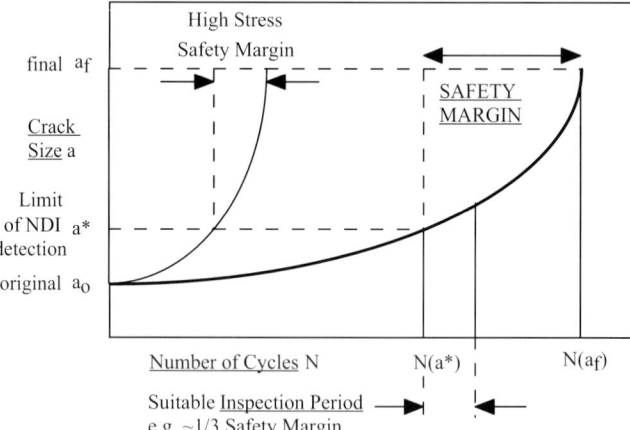

Fig. 4 Safety Margins and Subcritical Crack Growth. (The diagram shows schematically the growth of an initial defect under the cyclic stresses associated with the major take-off/landing cycles: see Ref. 9 for further detail and for a treatment of 'minor' cycles.)

Whereas the (2 MHz) ultrasonic resolution limit for *buried* defects in steel structures is of order 3 mm, the limit for 'highly reliable' detection of a *surface* crack in a gas turbine disc is defined as 0.75 mm. Gas turbine discs experience high stress amplitudes, as a result of the major take-off/landing cycles, and LEFM calculations show that, to achieve 10 000 flight cycles at a stress amplitude of 1000 MPa, the initial crack would have to be no greater than 0.15 mm in size. If a surface crack of this size were originally present, it should be detectable at the 'one-third life' inspection period and would be very obvious at the 'two-thirds life' inspection (see Ref. 9 for sample calculations and for a discussion of the treatment of 'minor', high-amplitude, cycles). The size of 0.15 mm (or $2a = 0.3$ mm for an embedded crack) is, however, very much smaller than conventional ultrasonic resolution limits. It may be technically possible to employ higher frequency (lower wavelength) ultrasound for extremely clean, relatively small, nickel discs, but there are many examples of other machine components whose fatigue lives are sustained by the control of the sizes of defects very much smaller than NDI limits. In such circumstances, control of defects rests on the quality and reliability of the processing and joining methods employed, although further protection may be given by surface heat-treatments or mechanical peening. The process/joining route must be proved first by exhaustive testing of small specimens. If these results are positive, it is possible to proceed to larger-scale pieces and then to full-size components. Under CAA regulations, it is necessary to justify the declaration of the fatigue life of a turbine disc in terms of the performance of full-size discs in spin-rigs. Due to the sensitivity of fatigue to surface condition, it is necessary to carry out tests on material machined in precisely the same manner as the service component and subjected to an identical surface treatment. Using results from the small test-pieces, fracture mechanics analyses are of value in assessing the potential viability of a scale-up test programme. This represents a substantial financial investment, which is worthwhile if it leads to a successful declaration of fatigue life, but is a serious loss, in terms of both testing costs and set-back to production plans, if the final test results do not match earlier promises.

A variety of defects can be present in engineering materials: shrinkage cavities or entrained oxide films in castings; refractory inclusions in powder-formed alloys; surface laps, machining marks, corrosion pits in wrought alloys; internal pores or brittle glassy phases in sintered engineering ceramics; a range of cracks in welded or riveted joints; regions of poor wetting in adhesively-bonded joints. Unexpected abuse in service, e.g. as a result of an impact, can also produce defects part-way through life. It is possible to examine failed test-pieces or the occasional failed service component to identify the nature of any defect from which the crack has grown. The primary technique is scanning electron microscopy (SEM) which combines high resolution with large depth-of-field. The chemical composition of any second-phase particle can be determined by analysing the X-rays generated by the electron probe. Surface embrittlement can be detected using Auger electron spectroscopy (AES) or secondary-ion mass-spectroscopy (SIMS). Local crystallographic orientations are analysed using the orientation imaging microscope (OIM). Since the aim is to identify the factors that caused the part to end its 'life' at the particular time that it did, the total set of examinations might be referred to as *forensic fractography*. Even at the observational level, the fractography is important. First, it may be used to distinguish between the 'normal' defects and their size-distribution (which control the mean and standard deviation of the data-base *S–N* curve) and any 'rogue' defects, which may have been responsible for a shorter-than-expected fatigue life. This information can be used to improve the quality-control on processing and fabrication. Second, comparisons can be made between full-size components and smaller testpieces to ensure that the causes of failure in the small samples are properly representative of the full-size part.

In machine components, two factors influence the fatigue life at a given stress-level: the *defect size* and the *residual stress distribution* produced by surface treatment. The *combination* of the two may be assessed experimentally by testing 'big blocks' which have received treatment identical to that in service components and employing fractography to measure and count fatigue striations. The inverse of the crack growth-rate is used to 'back-track' by integration to *the effective initial flaw size*, EIFS. This value is, in turn, used to calculate the fatigue life of a full-size component such as a disc. This is a *coupled* solution, suitable for making the decisions with respect to scale-up to full-size component testing, but *not* capable of providing *de-coupled* information, which would be needed to carry out *sensitivity* studies to decide whether better performance would be achieved, either by more stringent control of defect size or by inducing higher levels of compressive stress. The 'negative *K*' approach referred to in the previous section holds promise with respect to such 'de-coupling' but there are caveats concerning the monotonic nature of the residual stress *vis-à-vis* the mean stress in the fatigue cycle and also with the possible decrease in compressive stress level arising from stress-relaxation during service.

MATERIALS PROPERTIES

It is clear that many features of a material's composition, impurity content, processing route, fabrication procedure and surface treatment are critically important with respect to structural integrity issues. Here, attention is focused simply on 'material properties': more specifically, fracture toughness and fatigue-crack propagation-rate. There are many reviews giving detailed models of these properties (e.g. Refs 8, 10 and 11), so that only a summary is given in this section, concentrating on the reasons for scatter in values, trying to underpin statistical treatments based on examination of data with physical models of the fracture and fatigue mechanisms. In engineering alloys, under monotonic loading at low temperature in a non-interactive environment, there are basically four types of fracture process: *trans-granular cleavage* in steels and some other b.c.c. alloys; *inter-granular brittle fracture*, when the grain-boundary work-of-fracture has been lowered by the segregation of impurity elements; *trans-granular ductile fracture*, produced by the coalescence of inclusion-centred voids through inter-void plastic flow or via 'fast-shear' linkage; *inter-granular ductile fracture*, when there is a finely-spaced distribution of easily debondable particles situated on grain-boundaries, e.g. sulphides in 'over-heated' steels.

The brittle fracture modes, which are predominantly tensile-stress (propagation) controlled are modelled by the RKR/Curry:Knott[12,13] or other 'local approaches'.[14] When the load on a fracture-toughness test-piece is increased, the dislocations in the crack-tip plastic zone exert stress on brittle second-phase particles and can cause them to crack, to form potential brittle fracture nuclei. As a result of the high triaxial tensile stress-state in the plastic zone, high values of maximum tensile stress are generated, and when a new, 'virulent' micro-crack is nucleated *and* the tensile stress at a *critical distance* is sufficiently high, a critical condition is attained and fast fracture ensues. The Beremin local approach methodology[14] assigns a Weibull distribution to (an average of) the local tensile stress at fracture in a notched bar and computes the probability of this value being attained in the plastic zone. Extension of the Curry/Knott model to weld-metals[15] measured the sizes and locations of the fracture-initiating particles (oxide/silicate de-oxidation products) and produced validated links between fracture toughness and the microstructure/inclusion size-distribution. Assessment of effects of impurity-element segregation on intergranular brittle fracture follow a similar route. The critical condition is still the propagation of a microcrack nucleus, but segregation of impurity elements to grain-boundaries causes a progressive decrease in the work-of-fracture, which equates to continuing reductions in the critical values of local fracture stress required to propagate the micro-crack nucleus. This, in turn, implies that the critical event ahead of a yielded crack tip is met by progressively lower values of applied stress-intensity factor.

Early models for ductile fracture were based on the plastic expansion of a set of voids centred on non-metallic inclusions. The CTOD at initiation, δ_i, is then directly related to inclusion size and spacing, although the details of the linkage are also affected by the work-hardening rate.[16,17] If the inclusions are not uniformly spaced, local 'patches' of ductile extension may be observed along the crack front, although these tend to be smoothed out as the overall fracture grows.

In coarse-grained 'overheated' steels, there is a fine distribution of sulphides on prior austenite grain boundaries. On average the fatigue-crack tip in a *sharp-crack* fracture toughness test is initially located in the centre of a grain (although isolated segments may, by chance, intercept grain boundaries) and the value of δ_i is high because the crack tip is blunting in a rather clean matrix. Once the majority of the crack front intercepts boundaries, however, the crack propagates rapidly. The toughness exhibited by a test-piece having a large notch radius which samples many grain boundaries can be much lower than the sharp-crack result would suggest. Cavities on grain boundaries can also grow by (grain boundary) diffusion-controlled creep at high temperatures. Many structural alloys have rather widely-spaced non-metallic inclusions, but a high volume fraction of closely-spaced precipitates, which increase the flow stress by interfering with the movement of dislocations. In such alloys, the crack-tip ductility is limited, not by the simple coalescence of inclusion-centred voids, but by the initiation of a set of fine microvoids on the next coarsest set of particles: the coarser carbides in steels or dispersoids in aluminium alloys. These microvoids initiate when the strain in the matrix between the expanding large voids reaches a value sufficient to crack, shear or de-cohere the particles. This *fast shear* linkage shares some similarities with the brittle fracture processes. In A533B pressure-vessel steel, a coarse bainitic microstructure tends to crack and form microvoids much more easily than does a fine, tempered martensite. In a QT forging steel, it has been shown that the segregation which causes brittle fracture to occur in an intergranular manner at low temperature can also lower the work-of-fracture of carbide/matrix interfaces and hence enable fast-shear linkage to occur at a lower matrix strain and hence at a lower CTOD. Cold pre-strain reduces the local work-hardening rate in the matrix between voids and this also allows linkage to proceed at lower 'global' matrix strain.

Many of these features have relevance to what are normally regarded as 'fatigue' crack growth-rates.[8,18] Under cyclic loading, a crack propagates by whatever combinations of mechanism are able to operate. Even at room temperature, in a non-interactive environment, the 'classical' striation mode of propagation may be enhanced by *static modes* which respond to the value of K_{max} in the cycle. Such static modes include bursts of intergranular fracture in segregated QT forging steels, the onset of transgranular cleavage in plain carbon steels, the cracking of brittle silicon particles in aluminium alloy castings or silicon carbides in particulate aluminium matrix MMCs, the rapid growth of voids in 316 weld metals contrasted with parent plate, and the growth of microvoids around dispersoids in wrought structural aluminium alloys. The effects are most clearly observed at high ΔK (K_{max}) and high stress-ratio, but they may have contributed to values quoted in fatigue-crack growth-rate data-bases for 'material properties'.

The scatter associated with any particular mechanical property value is of importance because it forms part of the overall probabilistic failure assessment. Fracture toughness distributions may be fitted by a variety of statistical formulations, but, if these are not related to a physically-based model, there is concern that the tail of a distribution (to which the overall failure probability is highly sensitive) does not have the appropriate form. The modelling behind Local Approach/Weibull is based on the 'test volume' concept. A linear plastic zone dimension (e.g. a multiple of CTOD) is proportional to K^2 and so the process zone area is proportional to K^4 (as invoked also in the Curry/Knott treatment). If a weakest-link argument is accepted, the test volume is then proportional to the length of the crack front multiplied by K^4. For low probability events, the distribution needs to be a three-parameter Weibull, with a robust methodology for determining the *lower-bound* 'cut-off' value.[19] A two-parameter Weibull has a cut-off value of zero, so that its use for low-probability events is suspect. The *master curve* assumes a lower-bound of 20 MPam$^{0.5}$ for a range of pressure-vessel steels.[20] This single value may significantly underestimate the lower bound for some steels and possibly overestimate it for others.

Recent work has distinguished between *quasi-homogeneous* and *heterogeneous* materials.[21,22] No engineering material is truly homogeneous, but, if a steel contains a high volume fraction of relatively small carbides, distributed in a smooth 'well-behaved' fashion, a large number of potential crack nuclei will be sampled in the process-zone of every specimen in the batch tested. Whether the nucleus that produces the critical crack is the 97th, 99th or 98th percentile, the size of the particle in the 'well-behaved' distribution will not vary by any large amount and the critical fracture stress, which depends on the inverse square-root of the nucleus size, will be, to all intents and purposes, single-valued. It follows that the size of the process zone and the value of K at fracture will be the same for whichever specimen in the batch is tested. For such an ideal situation, the probability density function, pdf, would be a delta (spike-like) function and its integral, the cumulative distribution function, CDF, a step function. If random experimental errors are included, these, from the *central limit theorem*, are distributed normally and so the pdf becomes a *Gaussian* and the CDF the *error function*, erf. On normal probability paper, scaled about a mean of 50%, the error function plots as a straight line with 1 s.d. at 16% and 84%. Quasi-homogeneous behaviour would then correspond to a CDF which plotted as a straight line with 1 s.d. equal to the magnitude of random experimental errors. For carefully controlled tests on material of high pedigree, these might be set at ± 2 MPam$^{0.5}$ but examination of data for a number of engineering steels suggests a more 'pragmatic' value of ± 5 MPam$^{0.5}$. The proposal is that this value, in combination with a clear, tightly-populated, linear CDF defines quasi-homogeneous behaviour and that, if it is met, the data set can be used to extrapolate to a lower-bound value (at a given level of probability) with confidence.

The contrary example is found in the behaviour of deliberately-created *hetero-geneous* microstructures, composed of rather coarse spatial distributions of tough and brittle phases. Here, the fitting of data by simple distributions, such as the Weibull or the error function can lead to incredible lower-bound values (e.g. negative values of fracture toughness). It is important to use the suggested proposal (of a linear CDF with 1s.d. $< 5 \, \mathrm{MPam}^{0.5}$) *first* to decide whether or not extrapolation is permissible or whether forensic fractography needs to be used to separate-out bi-modality or multi-modality in the data. These data can be incorporated into an overall *weighted* dual (or multi-) distribution to denote *average* values e.g. in the transition range. If the arrestability of the cracks in the more brittle phase by the tougher phase were understood in detail, it might then prove possible to use this weighted distribution directly to estimate the lower-bound, but, at present, the safer approach is to model the more brittle phase microstructurally and to carry-out fracture toughness tests on this (quasi-homogeneous) microstructure. These data can then be used for extrapolation purposes. In many cases, it will not be possible to have access to materials or to have sufficient time and resource to perform the necessary fractography and testing to follow such a detailed sequence. It is, however, important to accept the general concept because it will lead to more intelligent 'expert judgment', to more confidence in the 'data-mining' of other existing data-bases and in identifying what is 'like-with-like'. The same principles apply to the interrogation of fatigue-crack growth-rate data, recognising the effects of 'static modes'. Some data may faithfully represent the particulars of the assessment being made: other data may be unrepresentative and the resulting assessment may err in either a conservative or a non-conservative manner.

Material properties may also change with time in service. Ignoring chemical interaction with the external environment (which is too large a field to treat in this paper) effects arise either as a result of 'abuse' or of the diffusion of atoms at elevated temperature. The service abuse might include work-hardening arising from an impact or other collision or the point-defect hardening of steels which occurs as a result of neutron irradiation in nuclear reactors.[23] Diffusion issues might possibly include hydrogen, which is mobile at room temperature, but, after welding and stress-relief, the hydrogen content is likely to increase only as a result of corrosion or external gas pressure. Carbon and, in particular, nitrogen atoms are quite mobile above room temperature and can produce *strain-ageing* in steels, which increases the yield strength and makes them more prone to brittle fracture (especially if a series of overloads is experienced). Phosphorus, tin, arsenic, antimony and sulphur are mobile in ferritic steels above 450°C under normal conditions and can produce intergranular embrittlement and carbide/matrix interface weakening. There are chemical reactors that operate at these tem-peratures and it is necessary to extract and test sample material at regular intervals. Lead appears to cause similar effects in 6xxx aluminium alloys at temperatures close to room temperature. In nuclear pressure-vessels, the neutrons impart energy to a primary knock-on atom, PKA, which then creates both individual point defects and defect clusters. The 'matrix damage' increases the yield strength and reduces work-hardening rates, so that *both* the cleavage fracture toughness is reduced *and* the 'upper shelf' ductile fracture toughness is reduced. In addition, the point defects allow copper precipitation-hardening at the service temperature of 300°C (rather than >450°C) and phosphorus segregation to grain-boundaries (again, normally >450°C). Both the increase in yield strength and the reduction in intergranular work-of-fracture render the material more susceptible to brittle fracture. The usual way of guarding against degradation of material properties resulting from service exposure is to embody *surveillance samples* in a structure. Sets of these can be removed at predetermined intervals and tested to assess property changes. Often, the samples are very much smaller in size than section thicknesses of concern and the implication of changes of properties in small-specimen geometries has to be considered with care. An important feature of any such assessments is the degree of understanding of local failure processes and how these are affected by the local values of stress, strain and triaxiality. A strong micro-mechanical input is required to underpin the use of surveillance data.

CLOSING COMMENTS

The early part of the paper described the factors leading to the successful design of a new venture or to a strong case for continued operation or life-extension. Reference was made to the HSE figure of a probability of 1 in 5000 per annum that a disaster involving 50 deaths might occur. Establishment of any overall failure probability requires the combination of separate probabilities. In safety-assessments, there are three main factors: stress, defect size and the material's resistance to crack extension. The discussion in the previous sections shows that, although there are still uncertainties in some aspects of 'stress' factors: such as resonance, thermal stress as a 'driving force', or stress-relaxation in force-controlled situations; it may be possible, as a first step, to assume that the stress 'dimension' is reasonably well understood and to focus on the variability in defect sizes and material properties.

The paper has made use of two main scenarios, which represent rather different issues with respect to structural integrity. One is that of the 'static' steel structure where the major threat to integrity is brittle fracture (or, possibly, plastic collapse). Here, the principles are: control of material *specification* (to give good toughness and to minimise inter-granular embrittlement); *quality-control* during fabrication; *ultrasonics* NDI (resolution $\sim$3 mm for buried defects); *operational codes*, drawn from long-term experience. Some of the main technical threats, in post World War II experience, have been lack-of-control during construction, occasional brittle fractures due to inadequate specification of material or fabrication method, transgression of 'best practice' rules *re* maintenance and operation (Three Mile Island, Chernobyl), and lack of appreciation of the service duty experienced at sea by tankers and bulkers (combined, perhaps, with non-ideal design, material or fabrication). In UK terms, 'best practice' is exemplified by the Sizewell B nuclear power station, particularly by the measures taken with respect to material specification and to NDI.

The second main area concerned aero-engine turbine discs. Here, the duty, in terms of stress levels and high cyclic stresses, is quite different from that of the steel structure. NDI techniques are generally *not* an option for routine defect control. For civil aircraft, the design life *has to be* >10 000 flight cycles to be economic. In extremely clean, well-controlled (expensive) nickel-base superalloys, the variations in fatigue-crack growth-rate are small and the fatigue life is dominated by the distribution of initial defects. These are, however, *coupled* with the residual stress distributions generated by shot-peening, so that the direct dependence on defect size is not easy to establish.

The two examples represent rather different approaches to the probability of failure. In the first, it is conceptually possible to take the probability that a given size of defect might be present in a structure and combine this with the probability that a defect of given size might produce failure. For many cases, the first probability has to be modified to the probability of detection of a defect of given size by the NDI technique used.[24] The discussions on extrapolation of fracture toughness data, analysis of fatigue-crack growth-rate data and the need for better informed expert judgment are highly relevant to the quality of these probabilistic assessments. For the aero-engine disc, the parameters change. The design parameter is fatigue-life; the crack-growth database is agreed; the issue is the initial defect distribution. For a life of 10 000 flight cycles or more at a stress-amplitude of 1000 MPa, the maximum size of initial defect that can be tolerated (making no allowance for residual stress) is 0.15 mm. Such a level can be effected by *process control* and the fatigue life established by *rigorous testing*, but the need for this and the sensitivity of the figure to the operational stress-amplitude needs to be appreciated throughout the whole design process. Particular issues concerning the timings of inspection periods have been discussed and there is the rather exciting possibility that, in the future, the fatigue database will be so well established that it may become feasible to make use of flight duty-spectra for individual aircraft, *either* to decide on the timing of the next inspection period *or* to 'juggle' flight schedules to fit in with pre-set inspections.

In both cases, *operational procedures* must be followed and routine *maintenance schedules* must be regularly carried-out. These procedures/schedules may also include advice on *condition monitoring*. The motorist is aware that if his or her car-engine begins to 'run rough' it is time to have it checked-over. There are similar indicators in engineering plant, which are often not given the full attention that they deserve, perhaps simply because there are no instructions in written form, or because the company's safety culture is not sufficiently well engrained at all levels. It may be necessary to offer positive (financial) rewards for the *reporting* of 'faults': too often, such reporting is tacitly discouraged. Excessive vibrations or clear evidence of leaks should indicate that not all is well with plant and carefully considered remedial action should be taken. It is salutary to recall that the Flixborough disaster occurred because there was an original leak in one of the reaction vessels. The action taken was not to explore the cause of this leak, but to remove the vessel and replace it with the 'jury rig' that failed. The vibrations on Galloping Gertie (the Tacoma Narrows bridge) in the months before final failure were so spectacular that it became a tourist attraction. In the light of such occurrences, the action taken to close the Millennium bridge across the Thames to fit stiffeners and dampers can only be viewed as prudent.

As sensor technology and signal processing continues to improve, there will be increasing opportunity to provide 'on-line' automatic *condition monitoring* throughout the service life of a structure or component. This is a separate exercise from that of periodic non-destructive testing, although, as experience is gained, it may be able to be used to help provide a more rational basis for the setting of inspection periods. At present, through-life sensor/actuator com-binations are commonly used to provide *operational* response to changes in conditions: whether a simple bimetallic strip to control temperature or a combination of accelerometers and actuators to trim an aircraft's wing when 'flying-by-wire'. Pressure sensors in the heavily glazed John Hancock building in Boston (Mass) are coupled to a floor of servo-hydraulic actuators, which are able to respond to damp down any effects of vibrations induced by wind forces. Such examples are focused on continued operation, rather than on the prediction of failure or remanent life, but there is no reason, in principle, why they should not feed back also into condition monitoring. The interrogation of sensors and the nature of feedback might be anticipated to be different for the 'static' steel structure and the aero-engine, in that a static structure may be quiescent for most of its life (i.e. any noise of noticeable amplitude is an 'event' requiring exploration), whereas an engine is continually generating a wide spectrum of vibrations and it is necessary to identify any significant change in an already noisy output. Many structures are, however, relatively pliant, deforming in response to the service duty, so that deflections and vibrations also need to be interrogated and responded to continuously.

Past experience of passive monitoring of elastic waves in structural components such as pressure vessels relates mainly to *acoustic emission*, with an array of piezo-electric (or magneto-strictive) transducers attached to the vessel's wall. Similar techniques are used in seismology, to detect earth tremors, and, in vulcanology, to monitor the activity of volcanoes, both active, (e.g. Etna) and dormant, (e.g. Vesuvius). There is great interest in the use of embedded fibre-optics containing Bragg gratings to measure service strains, for both civil engineering structures and composites in e.g. the masts of racing yachts. The aim in all cases is to anticipate and prevent catastrophe, but the nature of what is able to be done differs widely. The feedback from sensors in the Hancock building is used to provide active control via actuators, too large a loading on a bridge can be alleviated by traffic restrictions, too high a strain in a mast can be alleviated by relaxing the main-sheet or taking-in the sail, too high a level of acoustic emission in a pressure-vessel can be addressed by coming off load. In this last case, the method of alleviation may have consequences with respect to the economics of the process, and the question that is raised is 'what is the level (amplitude) of emission that gives cause for concern?' In the ideal case, this can be related to models for the growth of sub-critical cracks, or to the closeness of approach to fast fracture, but, even for reasonably well controlled tests in the laboratory, the underpinning science is, at best, semi-quantitative (see Ref. 25 for a fuller discussion). For volcanoes or earthquakes, there is little that can be done to alleviate the prime event, but sufficient prior warning enables people to be evacuated, or blocks set in place to try to divert lava flows. At the Seminar, slides of Etna were used to show the spread of lava from eruptions. It is tragic to note that, in mid-July, there was another major eruption. To quote from *The Times* (30 July 2001, p. 9), 'Bulldozers and prayers to the Virgin Mary were called into service yesterday in an increasingly desperate attempt to halt the lava flow . . .'

For aero-engines, Anuzis and Darby[26] have recently described the Rolls-Royce approach to engine health monitoring and diagnostics (EHM&D), which includes temperature monitoring, gas-path monitoring, oil systems diagnostics and vibration monitoring. Considering the last of these and using the analogy of the car engine, the basic questions are: (a) how is it determined that a part is 'running rough'?; (b) how 'rough' does the part have to run before mandatory remedial action is required? These questions define what an automated system has to do. The human ear is good at picking out those features of a noisy spectrum which indicate that an engine is not running as smoothly as it should be. An owner-driver becomes so used to the sound of his or her engine when running smoothly (over a range of r.p.m.), that the slightest deviation 'sets the alarm bells ringing'. For automated systems, it is important to ensure that the vibration sensors themselves do not add *variable* confusion to the noise spectrum, i.e. that they do not resonate at specific frequencies or saturate at different amplitudes at different frequencies. (There are concerns on this, stemming from acoustic emission experience). Given consistent sensor response, one promising approach which could address the first question is to make use of neural network analysis: 'training' a complete set of sensors on a smoothly-running system and then subjecting the same set of sensors to the system running in a more 'ragged' fashion. The sensors are then required to detect differences between the two sets of input vibration spectra (in terms of amplitudes, frequencies, spatial differences in source location etc.). A good analogy is that of the ear, trained in classical music, listening to a classical orchestral work. The trained ear is able to pick out a note played out of tune i.e. it can detect an incorrect basic frequency. It can identify the nature of the instrument on which the incorrect note is being played i.e. it can detect the particular combination of harmonics generated in the instrument (*timbre* or 'tone colour'). It can detect whether the note is played out of time, is too long or too short. It can locate the source spatially. That same ear, however, listening to Chinese or Indian music, would not necessarily be able to tell whether the note was, or was not, 'in tune' or 'in time', and may not be able to identify the source instrument, particularly if unfamiliar instruments are being played. The training aspect is clearly all-important to the detection of anomalies, but the training must be set in the context of the application. Once the 'running rough' anomaly has been detected, its significance can, at present, be assessed only experimentally: monitoring a component or near-identical set of components, and gradually increasing the severity of duty, whilst continuing to monitor output, (noting that the noise spectrum associated even with a smoothly-running part will change its amplitude/frequency characteristics as the part is forced to run under higher duty). Eventually, the aim is to provide models to relate changes in vibration spectra to specific types of events, but judging from the achievements in attempts to relate AE spectra to source events, assessment at present is likely to be empirically based. Even empirical correlations, backed by validations and calibration spectra, represent a positive step forward.

The examples of structural integrity assessments described in this paper are intended to set the overall subject in context. There is a clear path from *enterprise* to *plant*, to *function*, to *duty*, to the *resilience to bear duty without failure*. Each step in the path involves technical issues. The paper has described the scientific principles of relevance to these and drawn attention to areas where developments are possible. It is, however, vitally important to recognise that the successful operation of procedures relies heavily on 'human factors' and will succeed only if there is a safety-culture which is 'owned' and actively pursued by everyone, up to the highest levels of management in the company. Figure 2 is a much simplified visualisation of the design, lifetime, safety 'balancing act', but the factors that it represents need continuously to be borne in mind.

ACKNOWLEDGEMENTS

A part of the 'Current Concepts' content of this paper mirrors some of that in 'Structural Integrity – A Multi-Disciplinary Approach' presented by the author at the Conference 'Structural Integrity in the 21st Century' (Churchill College, Cambridge, 19–21 September 2000) and published in the Conference Proceedings (J. H. Edwards *et al.*, eds. EMAS, pp. 3–25). Thanks are due to the Conference organisers and to EMAS for permission to make use of this material.

REFERENCES

1. P. E. J. Flewitt and A. R. Dowling, 'The Development of Structural Integrity Assessment Methods: An Overview' in this volume.
2. *Reducing Risks, Protecting People*, Health and Safety Executive, 1999.
3. Nevile Shute, *No Highway*, William Heinemann, 1948.
4. J. F. Knott, 'Case Studies in Defence and Transportation Industries' in *Fracture and Fracture Mechanics – Case Studies*, R. B. Tait and G. G. Garrett eds, Pergamon, 1985, pp. 297–315.
5. P. Veron, C. A. Hippsley and J. F. Knott, 'Comparative Studies of Stress-relief Cracking in Relaxation Test Specimens and in a Full-scale Weldment' *Int. J. Pres. Ves. & Piping*, 1984, **16**, pp. 29–51.
6. P. Neumann, D. A. Miller and R. A. Ainsworth, 'Material Factors which Influence Remaining life of components Subject to Reheat Cracking' in *Fracture, Plastic Flow and Structural Integrity*, Peter Hirsch and David Lidbury eds, IOM Communications, 2000, pp. 175–184.
7. W. J. Jackson and J. C. Wright, 'Fracture Toughness Approach to Steel Castings Quality Assurance', *Metals Technology*, 1977, **4**, pp. 425–433.
8. J. F. Knott, 'Fatigue Design in Engineering Materials' in *Advances in Fracture Research* (Proc. 9th Intl. Conf. on Fracture) B. L. Karihaloo et al. eds, Pergamon, 1997, **3**, pp. 1213–1224.
9. J. F. Knott, 'The Durability of Rotating Components in Gas Turbines' in *Parsons 2000 – Advanced Materials for 21st Century Turbine and Power Plant*, A. Strang et al. eds, IOM Communications, 2000, pp. 950–960.
10. J. F. Knott, 'The Micro-Mechanisms of Fracture in Steels used for High Integrity Structural Components' in *Fracture, Plastic Flow and Structural Integrity*, Peter Hirsch and David Lidbury eds, IOM Communications, 2000, pp. 21–43.
11. J. F. Knott, 'Assessment of Fatigue in High-Duty Engineering Components' in *Reliability Assessment of Cyclically Loaded Engineering Structures*, R. A. Smith ed., Kluwer, 1997, pp. 137–164.
12. R. O. Ritchie, J. F. Knott and J. R. Rice, 'On the Relationship between Critical Tensile Stress and Fracture Toughness in Mild Steel', *Jnl. Mech. Phys. Solids*, 1973, **21**, pp. 395–410.

13. D. A. Curry and J. F. Knott, 'Effect of Microstructure on Cleavage Fracture Toughness of Quenched and Tempered Steels', *Metal Science*, 1979, **13**, pp. 341–346.
14. Beremin, 'A Local Criterion for Cleavage Fracture in Nuclear Pressure Vessel Steels' *Met. Trans.*, 1983, **14A**, pp. 2277–2287.
15. J. H. Tweed and J. F. Knott, 'Mechanisms of Failure in C–Mn Weld Metals', 1987, *Acta Metall.*, **35**, pp. 1401–1414.
16. J. F. Knott, 'Mechanisms of Fibrous Crack Extension in Engineering Alloys', *Metal Science*, 1980, **14**, pp. 327–336.
17. J. F. Knott, 'Effects of Microstructure and Stress State on Ductile Fracture in Metallic Alloys' in *Advances in Fracture Research* (Proc. 7th Intl. Conf. On Fracture), K. Salama et al. eds, Pergamon, 1989, **1**, pp. 125–138.
18. J. F. Knott, 'Models of Fatigue Crack Growth' in *Fatigue Crack Growth – 30 Years of Progress*, R. A. Smith ed., Pergamon, 1986, 31–52.
19. D. J. Neville and J. F. Knott, 'Fracture of Homogeneous and Inhomogeneous Materials', *Jnl. Mech. Phys. Solids*, 1986, **34**, pp. 243–291.
20. ASTM Standard E11921-98.
21. X. Zhang and J. F. Knott, 'Cleavage Fracture in Bainitic and Martensitic Micro-Structures', *Acta mater.*, 1999, **47**, pp. 3483–3495.
22. X. Zhang and J. F. Knott, 'The Statistical Modelling of Brittle Fracture in Homogeneous and Heterogeneous Micro-Structures', *Acta mater.*, 2000, **48**, pp. 2135–2146.
23. J. F. Knott and C. A. English, 'Views of TAGSI on the principles underlying the assessment of the mechanical properties of irradiated ferritic steel Reactor Pressure Vessels', *Intl. Jnl. of Pressure Vessels and Piping*, 1999, **76**, pp. 891–908.
24. *An Assessment of the Integrity of PWR Pressure Vessels* (Addendum to the Second Report of the Study Group, chaired since 1982 by Sir Peter Hirsch) UKAEA, 1987 (ISBN 0-7058-1155-7), Section 10.
25. G. Clark, D. J. H. Corderoy, N. W. Ringshall and J. F. Knott, 'Acoustic Emissions associated with Fracture Processes in Structural Steels', *Metal Science*, 1981, **15**, pp. 481–491.
26. P. Anuzis and E. C. Darby, 'A Forensic Approach to Condition Monitoring' presented at *Structural Integrity in the 21st Century*, Churchill College, Cambridge 19–21 September, 2000.